Classical Topics in Mathematics

Mathematics is the queen of sciences. She is pure, noble and attractive, and also has a distinct character in comparison with subjects in sciences such as physics: its permanent relevance and eternal validness of its theories and theorems. Whatever was once proved will stay true forever.

Mathematics is a vast subject, and many new concepts, theories and results spring up like mushrooms after spring rain. Similarly, there is also a large number of new mathematics books appearing in libraries and on bookshelves. Probably due to the usefulness of mathematics and its foundational nature, there seems to be more books in mathematics than in other subjects. On the other hand, only a limited number, or even a few, of them stand out and are appreciated and used by many people. The best test on the quality of books is the test of time.

In this series of books *Classical Topics in Mathematics*, we have selected books written by leading experts on topics which are well-tested by time. We believe that in spite of the passage of time, their power and value have not diminished, and hence they bear the trademark of the classical mathematics.

The first volumes of this series consist of an annotated version of Klein's masterpiece *Lectures on the icosahedron and the solution of equations of the fifth degree*, and the first English translation of Klein and Fricke's four big volumes on modular functions and automorphic functions. For this series, we have tried to pick books which share or reflect Klein's vision of the grand unity of mathematics.

The publication of this series of books is consistent with the motto of the Higher Education Press: to provide high quality books on the essential mathematics to the world mathematics community at an affordable price.

Classical Topics in Mathematics

(*Series Editor: Lizhen Ji*)

1. Lectures on the Theory of Elliptic Modular Functions, First Volume
Felix Klein, Robert Fricke (Authors), Arthur M. DuPre (Translator)

2. Lectures on the Theory of Elliptic Modular Functions, Second Volume
Felix Klein, Robert Fricke (Authors), Arthur M. DuPre (Translator)

3. Lectures on the Theory of Automorphic Functions, First Volume
Robert Fricke, Felix Klein (Authors), Arthur M. DuPre (Translator)

4. Lectures on the Theory of Automorphic Functions, Second Volume
Robert Fricke, Felix Klein (Authors), Arthur M. DuPre (Translator)

5. Lectures on the Icosahedron and the Solution of Equations of the Fifth Degree (With a New Introduction and Commentary)
Felix Klein, Peter Slodowy (Authors), George Gavin Morrice, Lei Yang (Translators)

6. The Bochner Technique in Differential Geometry
Hung-Hsi Wu

7. Lectures on Differential Equations and Differential Geometry
Louis Nirenberg

8. Notes on Jacquet-Langlands' Theory
Roger Godement

9. Kuga Varieties: Fiber Varieties over a Symmetric Space Whose Fibers Are Abelian Varieties
Michio Kuga

10. Arithmetic Groups and Reduction Theory
Armand Borel, Roger Godement, Carl Ludwig Siegel, André Weil (Authors), Lizhen Ji (Editor), Wolfgang Globke, Lizhen Ji, Enrico Leuzinger, Andreas Weber (Translators)

CTM 10

Classical Topics in Mathematics

Armand Borel
Roger Godement
Carl Ludwig Siegel
André Weil

Arithmetic Groups and Reduction Theory

算术群和约化理论

Edited by Lizhen Ji

Translated by Wolfgang Globke, Lizhen Ji, Enrico Leuzinger, Andreas Weber

HIGHER EDUCATION PRESS

Arithmetic Groups and Reduction Theory by Armand Borel, Roger Godement, Carl Ludwig Siegel, André Weil, first published 2020 by Higher Education Press

4 Dewai Dajie, Beijing 100120, P.R. China

Cataloging in Publication Data
图书在版编目（CIP）数据

算术群和约化理论 = Arithmetic Groups and Reduction Theory：英文 / 季理真主编. -- 北京：高等教育出版社，2020.6
（数学经典论题）
ISBN 978-7-04-053375-0

Ⅰ. ①算… Ⅱ. ①季… Ⅲ. ①算术子群 - 英文②约化理论 - 英文 Ⅳ. ① O152 ② O177.99

中国版本图书馆 CIP 数据核字（2020）第 025573 号

Copy Editor: Peng Li, Liping Wang
Cover Design: Zhi Zhang
Print Editor: Yimin Zhao

787mm×1092mm 1/16
9.25 Printed Sheets

Printed and Bound in P.R. China by Beijing Shengtong Printing Co.,Ltd.
ISBN: 978-7-04-053375-0
1 2020

Preface

This book in your hand consists of papers and lecture notes of four great mathematicians: Armand Borel, Roger Godement, Carl Ludwig Siegel and André Weil. Though they have different personalities, they share some common interests and good tastes, and they all write well. These writings in this book all deal with arithmetic subgroups of Lie groups and reduction theories for them, and reflect their deep knowledge of the subjects and their perspectives.

Before we explain the contents of these papers and the lecture notes, we first answer the question of why we publish these old writings, and why the reader should open and read this book.

Among these four authors mentioned above, probably the one with most opinions is Weil. In his autobiography titled *The apprenticeship of a mathematician*, Weil wrote:

> I had become convinced that what really counts in the history of humanity are the truly great minds, and the only way to get to know these minds was through direct contact with the works.

In his commentaries on the second volume of his collected works, Weil wrote:

> To comment on the works of Siegel has always appeared to me to be one of the tasks that a present-day mathematician may most usefully undertake.

As the reader can tell, the papers and lecture notes by Weil in this book are his expositions and expansions of some ideas and results of Siegel.

To understand the relations between these writings in this book better and to put them in a historical perspective, we recall the history of arithmetic groups and reduction theories.

As the name suggests, the reduction theory started with the reduction of binary quadratic forms. The earliest and best known binary quadratic form is $x^2 + y^2$ as in the Pythagorean theorem. Motivated by this, Fermat studied the problem of representing primes and asserted that prime numbers of the form $4m + 1$ can be written as sums of two squares, i.e., represented by the quadratic form $x^2 + y^2$. He also claimed several related results. Euler tried to prove them without complete success. Then in 1775, Lagrange developed a general theory of binary quadratic forms to prove these assertions of Fermat and more.

One key notion introduced by Lagrange is the equivalence relation between binary quadratic forms, and the basic problem is to find some canonical (or the simplest) representatives of each equivalence class. The notion of reduced forms was introduced as candidates for the simplest representatives. This turns out to be crucial to solving the problem on when an integer is represented by a quadratic form and also to deciding whether two quadratic forms are equivalent.

Though Legendre developed further the work of Lagrange on quadratic forms, the real breakthrough came with Gauss with the publication of his masterpiece *Disquisitiones Arithmeticae* in 1801. In this book, Gauss gave a systematic description of the theory of binary quadratic forms, and the role of the modular group $SL(2,\mathbb{Z})$ was made transparent.

The modular group $SL(2,\mathbb{Z})$ is the most basic example of arithmetic groups. Therefore, arithmetic groups and the reduction theory were developed hand in hand.

After the refinement of the work of Gauss by Dirichlet, Dedekind and Klein et al., the reduction theory of binary quadratic forms was reduced to understanding the action of $SL(2,\mathbb{Z})$ on the upper halfplane and its boundary, the real line.

Many other people also tried to generalize the theory of binary quadratic forms to quadratic forms of more than two variables, and Hermite was one of the most important contributors. He tied the reduction theory of positive definite quadratic forms to the problem on the minimum values taken by the quadratic forms on integral vectors and also developed the reduction of indefinite quadratic forms, building on his earlier work on the reduction of indefinite binary quadratic forms.

The next major figure was Minkowski. He initiated the theory of geometry of numbers to give a simple and transparent proof of the result of Hermite on the minimum value of a positive definite quadratic form on integral vectors and refined the definition of reduced positive definite quadratic forms. Minkowski gave a fundamental domain for the action of $SL(n,\mathbb{Z})$ on the space of positive definite quadratic forms of n variables. Then one important problem is to understand this Minkowski fundamental domain.

Siegel introduced the notion of Siegel sets for the group $SL(n,\mathbb{Z})$ acting on the space of quadratic forms and proved the Siegel finiteness property, which implies other finiteness results, for example, the finiteness of equivalence classes of indefinite integral quadratic forms in n variables of a fixed discriminant. Siegel discussed the reduction theory of positive definite quadratic forms systematically in a series of lectures in Japan, and the booklet based on the lecture notes is selected for this book.

One immediate problem is to generalize these results of Siegel for $SL(n,\mathbb{Z})$ to other classical groups. Siegel discussed the case of automorphism groups of involutive algebras. Weyl elaborated on these results of Siegel on two papers on the reduction theory.

Besides giving two Bourbaki seminar talks on Siegel's results, Weil also gave a course on discrete subgroups and reduction theory at the University of Chicago. All these writings of Weil are included in this book.

In spite of these efforts of Siegel, Weyl and Weil, there was still no general reduction theory for arithmetic subgroups of all semi-simple Lie groups, in particular, for exceptional Lie groups.

Motivated by the reduction of Hermite for indefinite quadratic forms, Borel and Harish-Chandra constructed coarse fundamental domains (called fundamental sets) for all arithmetic subgroups of semi-simple Lie groups in a seminal paper in 1962. As a consequence, they derived that such arithmetic subgroups are lattices of the Lie groups.

On the other hand, the fundamental sets constructed by Borel and Harish-Chandra are not defined in terms of intrinsic structures of the semi-simple Lie groups, and hence their shapes and structures at infinity cannot be described explicitly, as in the case of $SL(n)$ and other classical groups treated by Siegel, Weyl and Weil.

Then in 1962 Borel introduced the notion of generalized Siegel sets for all semi-simple Lie groups defined over rational numbers and showed that finite unions of Siegel sets give fundamental sets for arithmetic subgroups. He also explained applications of such a reduction theory for constructing compactifications of locally symmetric spaces. Borel first explained these results in some lectures, and its paper based on them is included in this book. Later he expanded some of the results in this paper into a book in 1969. But not everything in this paper is covered by the book. For example, compactifications of symmetric spaces were used in this paper to constructing fundamental domains, but not in the book.

One aspect of Borel's results on reduction theory is not so satisfactory. He defined a Siegel set for each rational parabolic subgroup. After showing that a finite union of Siegel sets forms a fundamental set for an arithmetic subgroup, he proved that there are only finitely many Γ-conjugacy classes of rational parabolic subgroups.

Borel told the editor many years ago, and others agree, that it seems more natural to prove first that there are only finitely many Γ-conjugacy classes of rational parabolic subgroups, then use this result to construct a finite union of Siegel sets as a fundamental set for the arithmetic subgroup Γ.

Godement and Weil showed that this approach works when the arithmetic subgroups are congruence subgroups by using the theory of adeles. These results are described by the Bourbaki talk of Godement included in this book.

From the brief discussion of the history of reduction theory above, it is clear that the papers and the lecture notes selected for this books give the eyewitness account of the reduction theory by some of the major players of the subject. Consequently, the papers and the lecture notes are arranged in the following order:

1. Siegel, *On the Reduction Theory of Quadratic Forms.*
2. Weil, *Reduction of Quadratic Forms, According to Minkowski and Siegel.*
3. Weil, *Groups of Indefinite Quadratic Forms and Alternating Bilinear Forms.*
4. Weil, *Discontinuous Subgroups of Classical Groups.*
5. Borel, *Fundamental Sets for Arithmetic Groups.*
6. Godement, *Fundamental Domains of Arithmetic Groups.*

The above discussion explained the motivation of arithmetic groups and the reduction theory from the theory of number theory, in particular, quadratic forms. Now we say a few words of the importance of arithmetic groups and applications of the reduction theory in geometry and topology.

First of all, arithmetic groups give an infinite family of explicit infinite groups in the geometric group theory and combinatorial group theory. At one point, the editor wrote a paper titled *What is the most beautiful group?*, and gave $SL(2,\mathbb{Z})$ as an answer. This group admits many generalizations. For example, besides the family $SL(n,\mathbb{Z})$, $n \geq 2$, the group $Sp(2g, \mathbb{Z})$ is another generalization and arises naturally when we consider the moduli spaces of Abelian varieties of dimension g and the mapping class group of compact oriented surfaces of genus g, which is one of the most basic objects in low dimensional topology.

To understand better algebraic properties of arithmetic groups, we need to study their actions on symmetric spaces. For this purpose, the reduction theory plays an important role. Besides finite generalization and presentation of arithmetic groups, the reduction theory was used to construct the Borel-Serre compactification of locally symmetric spaces which gives explicit models of classification spaces of arithmetic groups. The reduction theory also shows that arithmetic subgroups of semi-simple Lie groups are lattices. Consequently, quotients of symmetric spaces by arithmetic groups provide natural examples of locally symmetric spaces of finite volume. These spaces enjoy the celebrated Mostow strong rigidity, which motivated the Margulis superrigidity on irreducible lattices of higher rank semi-simple Lie groups, which implies the Margulis arithmeticity theorem for such lattices, which in turn show the special properties enjoyed by arithmetic groups.

Besides the modular interpretation of the locally symmetric spaces associated with $SL(n, \mathbb{Z})$ and $Sp(2g, \mathbb{Z})$, many other moduli spaces in number theory and algebraic geometry can also be identified with arithmetic locally symmetric spaces, especially the so-called Shimura varieties. The reduction theory plays a crucial role in understanding the geometry and compactifications (such as the Baily-Borel and the toroidal compactifications) of these spaces.

It is known that Gauss' theory of binary quadratic forms initiated the modern number theory, in particular, automorphic forms and automorphic representations. It will perhaps not be surprising that arithmetic groups and their reduction theory provide foundations of automorphic forms and automorphic representations.

Besides the lecture notes of Weil, all the papers selected for this book were not written in English. They have been translated into English for the first time for this book. The paper of Siegel was translated by Wolfgang Globke and Andreas Weber, and the paper of Godement by Enrico Leuzinger, and the papers of Borel and Weil by the editor.

I would like to thank Wolfgang Globke, Andreas Weber, and Enrico Leuzinger for their help, and Liping Wang for her support in this project.[1] Finally, I would like to thank Sylvie Weil for her permission for the inclusion of her father's lecture notes in this book.

Lizhen Ji
May 2018

[1] When this book was planned, Liping Wang was still at the Higher Education Press. Due to the long time this project took, she left HEP almost one year before I read the galley proof of this book. It is sad since this series of books was initiated by her. Hope that this series will continue its success.

Contents

1

On the Reduction Theory of Quadratic Forms*

Carl Ludwig Siegel

Translated by Wolfgang Globke and Andreas Weber

Preface

The fundamental domain of the group of units of an indefinite quadratic form with rational coefficients in n variables is not compact if this form is a null form[1], and it is well-known that this is always the case for $n > 4$. For this reason it seems interesting to compactify such fundamental domains in a suitable way by adding improper points. An important aspect of this is that the compactified domain should also have only finitely many image domains as neighbors. Due to the relationship with the theory of reduction of quadratic forms it seems helpful to first study the analogous problem for the domain of all reduced positive definite quadratic forms of given determinant.

The results allow some applications of geometric, arithmetic and analytic nature. For the well-known fundamental domain of the elliptic modular group, the parabolic point can be determined by a simple property of non-Euclidean lines. A corresponding characterization of the improper points for the fundamental domains of the unimodular group and the group of units is also conducted, where we consider the geodesic lines for the corresponding invariant Riemannian metric. Moreover, we will find that the representation of zero by an indefinite quadratic form is closely related to the different domains at infinity of the corresponding fundamental domain, namely, distinct cusps belong in a certain way to non-associated representations. Another application concerns the analytic theory of indefinite quadratic forms. There, it is necessary to prove the finiteness of the volume of the fundamental domains of the groups of units, as well as the convergence of certain other integrals associated to it. So far, this has been done in the literature in rather complicated ways, as a consequence of the integration variables not having been introduced in the most advantageous manner. In the

---

* Translated from *Zur Reduktionstheorie Quadratischer Formen*, The Mathematical Society of Japan, 1959.

1 *Translator's note:* This means isotropic in modern terminology.

following, the proof of finiteness will be given in a simplified manner that I occasionally used in lectures already. The coordinates used here arise from the compactification of the fundamental domain.

Let it be noted that the methods employed here can be transferred to analogous questions about involutory algebras over the field of rational numbers.

It has not escaped me that matrix computations can be replaced by vectors through appropriate manipulations, which gives the statements a certain geometric guise. As it has been said recently by evidently competent parties that the theory of quadratic forms is still in a chaotic state, it is perhaps appropriate to apologize for me still using an old-fashioned terminology and thus appear similar to M. Mourdain in Molière's comedy who was not even aware that he was speaking in prose. But I learned the now antiquated terminology four decades ago from Frobenius and Schur, in the meantime increased the chaos of this lingo myself, and am now speaking the way I learned it in my youth.

With regard to the bed of Procrustes in which many younger people have forced the beautiful body of Algebra, I may quote a few lines of a letter from Frobenius to Weber from the year 1893:

"Your announcement of a work on Algebra gives me great joy. Hopefully, you often follow the path of Dedekind, while avoiding some of the more abstract corners he likes to frequent nowadays. His latest edition contains so much beauty, the §173 is brilliant, but his permutations are disembodied, and it is unnecessary to drive abstraction that far. So I am almost glad that it is you who writes on Algebra and not our revered friend and master who was once contemplating the same thing."

I have lectured on the following subject matter at the "International Symposium on Algebraic Number Theory" at the universities of Kyoto, Nagoya, Osaka, Sendai, Tokyo.

Göttingen, 1958

Carl Ludwig Siegel

1 Normal Coordinates

Let $\boldsymbol{H} = (h_{kl})$ be a positive real symmetric matrix with n rows, whose diagonal elements are denoted by h_k. By successively completing the square one obtains the unique Jacobi normal form

$$\boldsymbol{H}[\boldsymbol{x}] = \sum_{k=1}^{n} t_k y_k^2, \qquad y_k = x_k + \sum_{l=k+1}^{n} d_{kl} x_l \tag{1}$$

with positive t_k and real d_{kl}, which can be expressed in the known manner as quotients of subdeterminants of the matrix $\boldsymbol{H}$. If $\boldsymbol{T}$ denotes the diagonal matrix formed by the t_k and $\boldsymbol{D}$ the triangular matrix formed by the d_{kl}, in which therefore $d_{kk} = 1\,(k = 1,\dots,n)$ and $d_{kl} = 0\,(1 \le l < k \le n)$, one obtains

$$\boldsymbol{H} = \boldsymbol{T}[\boldsymbol{D}], \qquad \boldsymbol{T} = [t_1,\dots,t_n], \qquad \boldsymbol{D} = (d_{kl}). \tag{2}$$

Conversely, if $\boldsymbol{T}$ is a diagonal matrix with positive diagonal elements t_k and $\boldsymbol{D}$ is a triangular matrix with real d_{kl}, then $\boldsymbol{H} > 0$ follows for the matrix defined by (2).

Since a unimodular transformation leaves the determinant of $\boldsymbol{H}$ invariant, we assume furthermore $|\boldsymbol{H}| = h$ with fixed positive h, so that also $|\boldsymbol{D}| = t_1 \cdots t_n = h$. With regard to a later applications, an arbitrary positive value of h is allowed. However, for the following analysis, one could normalize to $h = 1$ and extend by the mapping $\boldsymbol{H} \mapsto h^{\frac{1}{n}} \boldsymbol{H}$ to an arbitrary positive h later on.

We further introduce the $n-1$ ratios

$$u_k = \frac{t_k}{t_{k+1}} \quad (k = 1, \ldots, n-1),$$

which are all positive, and denote the $\frac{(n-1)(n+2)}{2}$ independent parameter $u_k, d_{kl} (k < l)$ as *normal coordinates* of $\boldsymbol{H}$. It is clear that due to the constraint $|\boldsymbol{H}| = h$, the elements h_{kl} of $\boldsymbol{H}$ are determined bijectively and continuously by the normal coordinates. The space of these matrices $\boldsymbol{H}$ is denoted by P. Let now $\overline{P}$ be the closure of P with respect to normal coordinates, i.e. the space that is obtained by demanding $u_k \geq 0 \, (k = 1, \ldots, n-1)$ with arbitrary real $d_{kl} \, (1 \leq k < l \leq n)$. One has to keep in mind that a point of $\overline{P} - P$ does not correspond to a matrix $\boldsymbol{H}$ anymore. Moreover, for any $\rho > 0$, $\overline{P}(\rho)$ denotes the part of $\overline{P}$ that is defined by the inequalities

$$0 \leq u_k \leq \rho, \qquad -\rho \leq d_{kl} \leq \rho \quad (1 \leq k < l \leq n). \tag{3}$$

Since this subset is bounded and closed it is compact.

The points of $\overline{P} - P$ are called *boundary points* and will be denoted by small Greek letters $\alpha, \beta, \ldots$. If a sequence $\boldsymbol{H}$ tends to a boundary point α, this will be expressed by $\boldsymbol{H} \Rightarrow \alpha$. Again one has to keep in mind that this statement refers to the limit with respect to normal coordinates. Now the boundary points shall be classified in greater detail. If $u_k = 0$ for precisely $k = \kappa_1, \ldots, \kappa_{r-1}$, then the respective boundary point is called of *type* $\{\kappa\} = \{\kappa_1, \ldots, \kappa_{r-1}\}$. We have $1 \leq r \leq n$ and one can assume $0 < \kappa_1 < \kappa_2 < \cdots < \kappa_{r-1} < n$. We also put $\kappa_0 = 0$, $\kappa_r = n$, and $\kappa_p - \kappa_{p-1} = j_p \; (p = 1, \ldots, r)$. The points of P themselves are of *empty type* for which $r = 1$. We say that a type $\{\lambda\} = \{\lambda_1, \ldots, \lambda_{s-1}\}$ is contained in $\{\kappa\}$ if the numbers $\lambda_1, \ldots, \lambda_{s-1}$ form a subset of $\kappa_1, \ldots, \kappa_{r-1}$. Of course, the type $\{1, \ldots, n-1\}$ contains all other types. The boundary points of a given type $\{\kappa\}$ determine a connected set P_κ which is defined by the following $n-1$ conditions

$$u_k = 0 \quad (k = \kappa_1, \ldots, \kappa_{r-1}), \qquad u_k > 0 \quad (k \neq \kappa_1, \ldots, \kappa_{r-1}).$$

With $\overline{P}\{\kappa\}$ we denote the part of $\overline{P}$ that is defined by the inequalities

$$u_k \geq 0 \quad (k = \kappa_1, \ldots, \kappa_{r-1}), \qquad u_k > 0 \quad (k \neq \kappa_1, \ldots, \kappa_{r-1}).$$

Obviously, it consists of the points of all types contained in $\{\kappa\}$. On $\overline{P}\{\kappa\}$ we introduce new coordinates which will be called *normal coordinates of type* $\{\kappa\}$. For this purpose we generalize the Jacobi transformation (1) by

$$\boldsymbol{H}[\boldsymbol{x}] = \sum_{p=1}^{r} \boldsymbol{T}_p[\boldsymbol{y}_p], \qquad \boldsymbol{y}_p = \boldsymbol{x}_p + \sum_{q=p+1}^{r} \boldsymbol{D}_{pq} \boldsymbol{x}_q \tag{4}$$

where $\boldsymbol{x}_p \; (p = 1, \ldots, r)$ is the column formed by the x_k with $\kappa_{p-1} < k \leq \kappa_p$. We call (4) the *normal form of type* $\{\kappa\}$. To show existence and uniqueness of this normal form, we analogously write

$$\boldsymbol{H}[\boldsymbol{x}]=\sum_{p,q=1}^{r}\boldsymbol{x}'_p\boldsymbol{H}_{pq}\boldsymbol{x}_q,\qquad \boldsymbol{H}_{qp}=\boldsymbol{H}'_{pq},\qquad \boldsymbol{H}_{pp}=\boldsymbol{H}_p \tag{5}$$

and note that because of $\boldsymbol{H}>0$ also $\boldsymbol{H}_1>0$, and in particular $|\boldsymbol{H}_1|\neq 0$. With a generalization of successively completing the square it follows

$$\boldsymbol{H}[\boldsymbol{x}]-\boldsymbol{H}_1[\boldsymbol{y}_1]=\sum_{p,q=2}^{r}\boldsymbol{x}'_p\boldsymbol{H}^*_{pq}\boldsymbol{x}_q, \tag{6}$$

where

$$\boldsymbol{y}_1=\boldsymbol{x}_1+\boldsymbol{H}_1^{-1}\sum_{q=2}^{r}\boldsymbol{H}_{1q}\boldsymbol{x}_q$$

and

$$\boldsymbol{H}^*_{pq}=\boldsymbol{H}_{pq}-\boldsymbol{H}_{p1}\boldsymbol{H}_1^{-1}\boldsymbol{H}_{1q}\qquad (p,q=2,\dots,r).$$

If $\boldsymbol{x}_1$ is determined by the condition $\boldsymbol{y}_1=0$ when arbitrary $\boldsymbol{x}_2,\dots,\boldsymbol{x}_r$ are given, it becomes evident that the quadratic form on the right hand side of (6) is positive definite, too. By induction on r the claim about (4) follows and we get $\boldsymbol{T}_1=\boldsymbol{H}_1$ and $\boldsymbol{D}_{1q}=\boldsymbol{H}_1^{-1}\boldsymbol{H}_{1q}\,(q=2,\dots,r)$. Furthermore, one sees that all $\boldsymbol{T}_p$ for $p=1,\dots,r$ are positive definite. As a generalization of (2) we obtain with (4) the matrix formula

$$\boldsymbol{H}=\hat{\boldsymbol{T}}[\hat{\boldsymbol{D}}],\qquad \hat{\boldsymbol{T}}=[\boldsymbol{T}_1,\dots,\boldsymbol{T}_r],\qquad \hat{\boldsymbol{D}}=(\boldsymbol{D}_{pq}),$$

where $\hat{\boldsymbol{T}}$ is formed by the boxes $\boldsymbol{T}_1,\dots,\boldsymbol{T}_r$ along the diagonal, and, furthermore, the boxes $\boldsymbol{D}_{pq}\,(1\le p<q\le r)$ of $\hat{\boldsymbol{D}}$ appear in (4) while $\boldsymbol{D}_{pq}=0\,(1\le q<p\le r)$ and $\boldsymbol{D}_{pp}\,(p=1,\dots,r)$ is the identity matrix with j_p rows. We call $\hat{\boldsymbol{T}}$ a *diagonal matrix of type* $\{\kappa\}$ and $\hat{\boldsymbol{D}}$ a *triangular matrix of type* $\{\kappa\}$.

If one puts $|\boldsymbol{T}_p|=\tau_p^{j_p}$, $\tau_p>0$, where again j_p denotes the number of rows of $\boldsymbol{T}_p$, one obtains

$$\boldsymbol{T}_p=\tau_p\boldsymbol{R}_p,\qquad \boldsymbol{R}_p>0,\qquad |\boldsymbol{R}_p|=1\qquad (p=1,\dots,r).$$

At last, we set

$$v_p=\frac{\tau_p}{\tau_{p+1}}\qquad (p=1,\dots,r-1).$$

These v_p together with the elements of all $\boldsymbol{R}_p\,(p=1,\dots,r)$ and $\boldsymbol{D}_{pq}\,(1\le p<q\le r)$ are called *normal coordinates of type* $\{\kappa\}$. One needs to take into account that the elements of the positive symmetric matrix $\boldsymbol{R}_p$ are related by the condition $|\boldsymbol{R}_p|=1$ and that τ_r can be derived from the remaining τ_p by virtue of the equation

$$|\boldsymbol{T}_1|\cdots|\boldsymbol{T}_r|=|\boldsymbol{H}|=h.$$

Obviously, the original normal coordinates have type $\{1,\dots,n-1\}$. To obtain the relation between both kinds of coordinates, one applies the Jacobi transformation to $\boldsymbol{R}_p$, so that one obtains the decomposition

$$\boldsymbol{R}_p=\boldsymbol{S}_p[\boldsymbol{C}_p]\qquad (p=1,\dots,r)$$

with a diagonal matrix $\boldsymbol{S}_p$ and a triangular matrix $\boldsymbol{C}_p$. With (4) it follows that

$$\begin{aligned} \boldsymbol{H}[\boldsymbol{x}] &= \sum_{p=1}^{r} \tau_p \boldsymbol{S}_p[\boldsymbol{C}_p \boldsymbol{y}_p], \\ \boldsymbol{C}_p \boldsymbol{y}_p &= \boldsymbol{C}_p \boldsymbol{x}_p + \sum_{q=p+1}^{r} \boldsymbol{C}_p \boldsymbol{D}_{pq} \boldsymbol{x}_q \qquad (p = 1,\ldots,r). \end{aligned} \tag{7}$$

Because of the uniqueness of the Jacobi normal form, the matrix $\boldsymbol{T}$ there is composed of the boxes $\tau_p \boldsymbol{S}_p$ $(p = 1,\ldots,r)$ on the diagonal, whereas $\boldsymbol{D}$ is the matrix of the system of linear forms in $x_1,\ldots,x_n$ on the right hand side of (7). Hence, the diagonal elements of $\tau_p \boldsymbol{S}_p$ have the values t_k $(\kappa_{p-1} < k \le \kappa_p)$ and one gets

$$\tau_p^{j_p} = \prod_{k=\kappa_{p-1}+1}^{\kappa_p} t_k = t_{\kappa_p}^{j_p} \prod_{s=1}^{j_p-1} u_{s+\kappa_{p-1}}^{s} \qquad (p = 1,\ldots,r),$$

$$t_{\kappa_{p-1}} = t_{\kappa_p} \prod_{k=\kappa_{p-1}}^{\kappa_p-1} u_k \qquad (p = 2,\ldots,r).$$

From this it becomes evident that the fractions v_p / u_{κ_p} $(p = 1,\ldots,r-1)$ tend to a positive limit if the point $\boldsymbol{H}$ tends to a boundary point in $\overline{P}\{\kappa\}$. The same holds true for the diagonal elements of $\boldsymbol{S}_p$ $(p = 1,\ldots,r)$, i.e. for the fractions t_k/τ_p $(\kappa_{p-1} < k \le \kappa_p)$. In particular, all the v_p tend to 0 if and only if the boundary point is of type $\{\kappa\}$, too. Hence, the normal coordinates of type $\{\kappa\}$ are also defined for the boundary points on $\overline{P}\{\kappa\}$ and the variability domain is given by the conditions

$$v_p \ge 0 \quad (p = 1,\ldots,r-1), \qquad \boldsymbol{R}_p > 0 \quad (p = 1,\ldots,r), \qquad |\boldsymbol{R}_p| = 1$$

for arbitrary real $\boldsymbol{D}_{pq}$ $(1 \le p < q \le r)$. The relationship with the original normal coordinates is bijective and continuous, also in the boundary points.

2 Linear coordinates

For a later purpose it is useful to introduce in an obvious manner another kind of coordinates that is also assigned to type $\{\kappa\}$. Namely, one decomposes the matrix $\boldsymbol{H}$ as in (5) in r^2 boxes $\boldsymbol{H}_{pq}$ and defines

$$|\boldsymbol{H}_p| = \omega_p^{j_p}, \qquad \omega_p > 0, \qquad \boldsymbol{H}_{pq} = \omega_p \boldsymbol{F}_{pq}, \qquad \boldsymbol{F}_{pp} = \boldsymbol{F}_p \qquad (1 \le p \le q \le r), \tag{8}$$

such that $\boldsymbol{F}_p > 0$ and $|\boldsymbol{F}_p| = 1$ and furthermore,

$$\frac{\omega_p}{\omega_{p+1}} = w_p \qquad (p = 1,\ldots,r-1).$$

The $r-1$ positive parameters w_p together with all elements of all matrices $\boldsymbol{F}_{pq}$ $(1 \le p \le q \le r)$ are called *linear coordinates of type* $\{\kappa\}$. It is evident that the elements of the matrix $\omega_r^{-1}\boldsymbol{H}$ are polynomials in the linear coordinates; ω_r^{-n} is also such a polynomial because of $|\boldsymbol{H}| = h$. The transformation from normal coordinates to linear coordinates of the same type follows from the comparison of (4) with (5), which yields the formulas

$$\begin{aligned} \boldsymbol{H}_p &= \boldsymbol{T}_p + \sum_{s=1}^{p-1} \boldsymbol{T}_s[\boldsymbol{D}_{sp}] \qquad (p = 1,\dots,r), \\ \boldsymbol{H}_{pq} &= \boldsymbol{T}_p\boldsymbol{D}_{pq} + \sum_{s=1}^{p-1} \boldsymbol{D}'_{sp}\boldsymbol{T}_s\boldsymbol{D}_{sq} \qquad (1 \le p < q \le r). \end{aligned}$$

Hence,

$$\begin{aligned} \frac{\omega_p}{\tau_p}\boldsymbol{F}_p &= \boldsymbol{R}_p + \sum_{s=1}^{p-1} \frac{\tau_s}{\tau_p}\boldsymbol{R}_s[\boldsymbol{D}_{sp}], \\ \frac{\omega_p}{\tau_p}\boldsymbol{F}_{pq} &= \boldsymbol{R}_p\boldsymbol{D}_{pq} + \sum_{s=1}^{p-1} \frac{\tau_s}{\tau_p}\boldsymbol{D}'_{sp}\boldsymbol{R}_s\boldsymbol{D}_{sq}. \end{aligned} \tag{9}$$

From this one obtains $(\omega_p/\tau_p)^{j_p}$ $(p = 1,\dots,r)$ as a polynomial in the normal coordinates of type $\{\kappa\}$ that is ≥ 1 on all of $\overline{P}\{\kappa\}$; therefore the elements of the matrices $\boldsymbol{F}_p$ and $\boldsymbol{F}_{pq}$ are also continuous functions there with respect to normal coordinates. Furthermore, $(w_p/v_p)^{j_p j_{p+1}}$ is a continuous rational function there for $p = 1,\dots,r-1$, hence the fraction w_p/v_p is also continuous and $w_p \to 0$ only if $v_p \to 0$. Thus, the linear coordinates are defined for the boundary points on $\overline{P}\{\kappa\}$. Once it is proven that the mapping is bijective, it follows that the normal coordinates are continuous as functions of the linear coordinates.

To prove this, we begin with the formulas

$$\begin{aligned} \frac{\tau_p}{\omega_p}\boldsymbol{R}_p &= \boldsymbol{F}_p - \sum_{s=1}^{p-1} \frac{\omega_s}{\omega_p}\frac{\tau_s}{\omega_s}\boldsymbol{R}_s[\boldsymbol{D}_{sp}], \\ \frac{\tau_p}{\omega_p}\boldsymbol{R}_s\boldsymbol{D}_{pq} &= \boldsymbol{F}_{pq} - \sum_{s=1}^{p-1} \frac{\omega_s}{\omega_p}\frac{\tau_s}{\omega_s}\boldsymbol{D}'_{sp}\boldsymbol{R}_s\boldsymbol{D}_{sq} \end{aligned}$$

and assume that τ_p/ω_p, $\boldsymbol{R}_p$, $\boldsymbol{D}_{pq}$ are uniquely determined by the linear coordinates if $p = 1,\dots,m-1$ and $q = m,\dots,r$, where $1 \le m \le r$ and the assumption is empty if $m = 1$. The formulas with $p = m$ yield uniquely and successively $(\tau_m/\omega_m)^{j_m}$, τ_m/ω_m, $\boldsymbol{R}_m$, and $\boldsymbol{D}_{mq}$ $(q = m+1,\dots,r)$, which completes the induction and proves the assertion. It is possible to give explicit formulas for the inverse transformation using determinants. However, for what follows, this is not needed.

Because of the proven continuous dependency one can choose normal coordinates or linear coordinates of the respective type $\{\kappa\}$ for the determination of the neighborhood of boundary points. More generally, one can use coordinates of any type that contains $\{\kappa\}$. In particular, we let $\boldsymbol{H}$ tend to a boundary point of type $\{\kappa\}$ and use coordinates of this type. In the limiting process all v_p tend to 0 and from (9) it follows that $\omega_p/\tau_p \to 1$. In the boundary point itself we have

$$\boldsymbol{F}_p = \boldsymbol{R}_p, \qquad \boldsymbol{F}_{pq} = \boldsymbol{F}_p\boldsymbol{D}_{pq} \qquad (1 \le p \le q \le r), \tag{10}$$

and furthermore $w_p = v_p = 0\,(p = 1,\dots,r-1)$. These are the simple relations between both kinds of coordinates for the boundary points of the same type.

3 Compactification of the Minkowski domain

With the help of normal coordinates, it is easy to compactify the domain of reduced positive definite quadratic forms $\boldsymbol{H}[\boldsymbol{x}]$. Let $|\boldsymbol{H}| = h$ and $\boldsymbol{H}$ be reduced in the sense of Hermite and Minkowski. It is known that there is a positive number c, only depending on n, such that $\boldsymbol{H}$ is contained in $\overline{P}(c)$. This domain however is compact. For the desired compactification one only has to add the boundary points that can be reached by a convergent sequence in the Minkowski domain.

Now one needs to determine the respective boundary points. To do this, one needs to take a closer look at the reduction conditions. These are all homogeneous linear inequalities for the elements h_{kl} of $\boldsymbol{H}$:

$$h_{kl} \geq 0 \qquad (l = k+1 = 2, \ldots, n) \tag{11}$$

and

$$\boldsymbol{H}[\boldsymbol{g}] \geq h_k \qquad (k = 1, \ldots, n) \tag{12}$$

for every column $\boldsymbol{g}$ consisting of n integer numbers $g_1, \ldots, g_n$ with greatest common divisor $(g_1, \ldots, g_n) = 1$. In case $\pm\boldsymbol{g}$ is the k-th column of the identity matrix, (12) is fulfilled with equality. We omit these trivially fulfilled conditions. Then the remaining conditions are not fulfilled with equality for $\boldsymbol{H}$. It is known that the infinitely many reduction conditions (11) and (12) contain a fixed finite system that implies all the others. Furthermore, together with the condition $|\boldsymbol{H}| = h > 0$, it follows from (12) that $\boldsymbol{H} > 0$, i.e. $\boldsymbol{H}$ is contained in P. Let $R_h = R$ denote the domain in the determinant level surface $|\boldsymbol{H}| = h$ that is defined by the inequalities (11) and (12).

Let now $\boldsymbol{H}$ converge on R to a boundary point of type $\{\kappa\}$. Later on, it turns out that for each type there exist such boundary points of R. We introduce linear coordinates of type $\{\kappa\}$ and denote by $\hat{\boldsymbol{F}}_p, \hat{\boldsymbol{F}}_{pq}$ the limits of $\boldsymbol{F}_p, \boldsymbol{F}_{pq}$. Because of $\boldsymbol{F}_p > 0$, $|\boldsymbol{F}_p| = 1$ we also have $\hat{\boldsymbol{F}}_p > 0$, $|\hat{\boldsymbol{F}}_p| = 1$. From (5), (8), (11) it follows firstly that the corresponding reduction conditions (11) are satisfied for $\hat{\boldsymbol{F}}_p\,(p = 1, \ldots, r)$ instead of $\boldsymbol{H}$, and also that the last element of the first column of $\hat{\boldsymbol{F}}_{pq}\,(q = p+1 = 2, \ldots, r)$ is non-negative. In order to discuss the behavior of the reduction conditions (12) in the limit process, one decomposes the column $\boldsymbol{g}$ into the parts $\boldsymbol{g}_p\,(p = 1, \ldots, r)$ and sets for fixed p initially $\boldsymbol{g}_m = 0\,(m \neq p)$, while with given k $(\kappa_{p-1} < k \leq \kappa_p)$ the condition $(g_k, \ldots, g_{\kappa_p}) = 1$ is fulfilled for the elements of $\boldsymbol{g}_p$. It then follows from (5), (8), (12) together with what was shown before that $\hat{\boldsymbol{F}}_p$ is reduced.

To obtain further conclusions about $\hat{\boldsymbol{F}}_{pq}\,(p < q)$ one chooses an arbitrary $\boldsymbol{g}_p \neq 0$, further $\boldsymbol{g}_q = \boldsymbol{e}_q$, where $\boldsymbol{e}_q$ is a column of the identity matrix with j_q rows, and otherwise $\boldsymbol{g}_m = 0\,(m \neq p, q)$. From (5), (8), (12) it follows

$$\omega_p \boldsymbol{g}_p'(\boldsymbol{F}_p\boldsymbol{g}_p + 2\boldsymbol{F}_{pq}\boldsymbol{e}_q) + \omega_q \boldsymbol{F}_q[\boldsymbol{e}_q] \geq \omega_q \boldsymbol{F}_q[\boldsymbol{e}_q]$$

and hence,

$$\boldsymbol{F}_p[\boldsymbol{g}_p] + 2\boldsymbol{g}_p'\boldsymbol{F}_{pq}\boldsymbol{e}_q \geq 0 \qquad (1 \leq p < q \leq r). \tag{13}$$

If one introduces normal coordinates $\hat{\boldsymbol{R}}_p, \hat{\boldsymbol{D}}_{pq}$ for the boundary point, then it follows from (13) together with (10) for the limit that the inequalities

$$\hat{\boldsymbol{R}}_p[\boldsymbol{g}_p + \hat{\boldsymbol{D}}_{pq}\boldsymbol{e}_q] \geq \hat{\boldsymbol{R}}_p[\hat{\boldsymbol{D}}_{pq}\boldsymbol{e}_q]$$

hold true for all integral columns $\boldsymbol{g}_p$. This is equivalent to the conditions

$$\hat{\boldsymbol{R}}_p[\boldsymbol{g}_p + \boldsymbol{d}_p] \geq \hat{\boldsymbol{R}}_p[\boldsymbol{d}_p] \qquad (p = 1, \ldots, r-1) \tag{14}$$

for all columns $\boldsymbol{d}_p$ of the matrices $\hat{\boldsymbol{D}}_{pq}$ $(q = p+1, \ldots, r)$.

Before we continue the analysis of the boundary points of R we want to discuss the conditions (14) independently from the antecedent.

4 The translation group

In n-dimensional Euclidean space Ω let Λ denote a translation group generated by n independent vectors, i.e. the elements of Λ form a lattice. One obtains a fundamental domain F in Ω for Λ if one chooses in every class of Λ-equivalent vectors one of smallest length. If one chooses the generating elements of Λ as unit vectors, this fundamental domain is defined by the inequality

$$\boldsymbol{H}[\boldsymbol{g} + \boldsymbol{x}] \geq \boldsymbol{H}[\boldsymbol{x}], \tag{15}$$

where $\boldsymbol{H}[\boldsymbol{x}]$ denotes a positive definite quadratic form and one can again assume $|\boldsymbol{H}| = h$. In (15) one has to allow all integral columns for $\boldsymbol{g}$ and one can exclude the trivial case $\boldsymbol{g} = 0$. Furthermore, if the basis of the lattice is chosen appropriately, then $\boldsymbol{H}$ is reduced. Thus, the connection to (14) is made.

Since $\boldsymbol{x}$ and also $-\boldsymbol{x}$ satisfy the inequalities (15), $\boldsymbol{x} = 0$ is the center of the fundamental domain $F = F(\boldsymbol{H})$. Moreover, it follows immediately from the conditions

$$\boldsymbol{H}[\boldsymbol{g}] + 2\boldsymbol{g}'\boldsymbol{H}\boldsymbol{x} \geq 0, \tag{16}$$

which are equivalent to (15), that F is convex. It is plausible that F is bounded by finitely many planes, i.e. F is a convex polyhedron with center at the point of origin. Furthermore, we will show that there is, independent of $\boldsymbol{H}$, a fixed finite set of integral columns $\boldsymbol{g}$ such that the respective finitely many inequalities (15) imply all the others. To prove this, one takes the matrix

$$\boldsymbol{H}_0 = \begin{pmatrix} \boldsymbol{H} & \boldsymbol{z} \\ \boldsymbol{z}' & \lambda \end{pmatrix}$$

with an arbitrary column $\boldsymbol{z}$ and a scalar λ that satisfies the condition

$$\lambda > \boldsymbol{H}^{-1}[\boldsymbol{z}] + h_n, \tag{17}$$

where h_n again denotes the last diagonal element of $\boldsymbol{H}$. Since $\boldsymbol{H}$ is reduced, we have $h_k \leq h_n$ $(k = 1, \ldots, n)$. We now analyze under which conditions $\boldsymbol{H}_0$ is reduced, too. The inequalities (11) imply that for $\boldsymbol{H}_0$ the last element of $\boldsymbol{z}$ must be non-negative. From the inequalities (12) it follows

$$\boldsymbol{H}[\boldsymbol{g}] + 2\boldsymbol{g}'\boldsymbol{z}g + \lambda g^2 \geq h_k \qquad (k = 1, \ldots, n) \tag{18}$$

for $(g_k,\dots,g_n,g)=1$ and

$$\boldsymbol{H}[\boldsymbol{g}]+2\boldsymbol{g}'\boldsymbol{z}+\lambda\geq\lambda \tag{19}$$

for arbitrary integral $\boldsymbol{g}$. If one puts $\boldsymbol{x}=\boldsymbol{H}^{-1}\boldsymbol{z}$, (19) is equivalent to (16). Since $\boldsymbol{H}$ is reduced, (18) is fulfilled for $g=0$. On the other hand, if $g\neq 0$, i.e. $g^2\geq 1$, because of (17) and because of $\boldsymbol{H}[\boldsymbol{x}]=\boldsymbol{H}^{-1}[\boldsymbol{z}]$ the left hand side of (18) becomes

$$\boldsymbol{H}[\boldsymbol{g}+\boldsymbol{x}g]+(\lambda-\boldsymbol{H}[\boldsymbol{x}])g^2\geq h_n g^2\geq h_k, \tag{20}$$

i.e. (18) is satisfied. It follows that $\boldsymbol{H}_0$ is reduced if and only if $\boldsymbol{x}$ lies in $F(\boldsymbol{H})$ and the last element of $\boldsymbol{z}$ is non-negative. Since all reduction conditions for $\boldsymbol{H}_0$ follow from a certain finite number of conditions, one sees that all the inequalities (15) follow from a certain finite number of inequalities among them. The form of the left hand side of (20) also shows that the elements $x_1,\dots,x_n$ of $\boldsymbol{x}$ appear as normal coordinates of type $\{n\}$ for the matrix $\boldsymbol{H}_0$. Thus, $F(\boldsymbol{H})$ is uniformly bounded with respect to $\boldsymbol{H}$, and, because of (16), the boundary planes of $F(\boldsymbol{H})$ change continuously with $\boldsymbol{H}$.

The points $\boldsymbol{x}$ in $F(\boldsymbol{H})$ are called *reduced with respect to* $\boldsymbol{H}$. $\boldsymbol{x}$ and also $-\boldsymbol{x}$ are reduced with respect to $\boldsymbol{H}$. If one adds to the translations $\boldsymbol{x}\mapsto\boldsymbol{x}+\boldsymbol{g}$ also the reflections $\boldsymbol{x}\mapsto\boldsymbol{g}-\boldsymbol{x}$, then for the group Λ^* extended in this way one obtains a fundamental domain $F^*=F^*(\boldsymbol{H})$ by bisecting $F(\boldsymbol{H})$ using a plane through $\boldsymbol{x}=0$. This may be achieved by the condition that the last element of $\boldsymbol{H}\boldsymbol{x}=\boldsymbol{z}$ is non-negative, i.e.

$$\boldsymbol{e}_n'\boldsymbol{H}\boldsymbol{x}\geq 0,$$

where $\boldsymbol{e}_n$ denotes the last column of the identity matrix. The points of $F^*(\boldsymbol{H})$, defined in this way, are called *tightly reduced with respect to* $\boldsymbol{H}$. In case of $\boldsymbol{H}=\boldsymbol{E}$, or more generally if $\boldsymbol{H}$ is a diagonal matrix, F is the cube $-\frac{1}{2}\leq x_k\leq\frac{1}{2}$ $(k=1,\dots,n)$ and F^* is the half-cube $-\frac{1}{2}\leq x_k\leq\frac{1}{2}$ $(k=1,\dots,n-1)$, $0\leq x_n\leq\frac{1}{2}$.

5 Reduced boundary points

As in Section 3 we take a sequence of points $\boldsymbol{H}$ in R that converges to a boundary point of type $\{\kappa\}$ with linear coordinates $\hat{\boldsymbol{F}}_p,\hat{\boldsymbol{F}}_{pq}$. For the respective normal coordinates $\hat{\boldsymbol{R}}_p=\hat{\boldsymbol{F}}_p$, $\hat{\boldsymbol{D}}_{pq}=\hat{\boldsymbol{F}}_p^{-1}\hat{\boldsymbol{F}}_{pq}$, the conditions (14) imply that all columns of the matrices $\hat{\boldsymbol{D}}_{pq}$ $(q=p+1,\dots,r)$ are reduced with respect to $\hat{\boldsymbol{R}}_p$ $(p=1,\dots,r-1)$. Furthermore, the first column of $\hat{\boldsymbol{D}}_{pq}$ $(q=p+1,\dots,r)$ is even tightly reduced with respect to $\hat{\boldsymbol{R}}_p$ $(p=1,\dots,r-1)$. Moreover, it was already stated that all $\hat{\boldsymbol{R}}_p$ $(p=1,\dots,r)$ are reduced in the sense of Minkowski. A boundary point of type $\{\kappa\}$ is now called *reduced* if its normal coordinates of type $\{\kappa\}$ satisfy all the aforementioned reduction conditions. Conversely, it will now be shown that each reduced boundary point can be reached by a convergent sequence in R.

All reduction conditions follow from a fixed system $\Theta=\Theta\{\kappa\}$ of finitely many among them, and these are homogeneous linear inequalities for the elements of the linear coordinates $\hat{\boldsymbol{F}}_p,\hat{\boldsymbol{F}}_{pq}$. Firstly, we take a look at a boundary point α for which all inequalities of the system Θ are strictly satisfied, i.e. with exclusion of the equality sign.

By continuity it then follows that α satisfies strictly all of the infinitely many reduction conditions.

We now prove that all points of P that are contained in a sufficiently small neighborhood of α in $\overline{P}\{\kappa\}$ lie in the interior of R. If this was false, there would be a sequence $\boldsymbol{H} \Rightarrow \alpha$ for which a fixed one among the conditions (11) and (12) is not strictly satisfied. If this is a condition (11), when taking the limit, $h_{kl} \leq 0$ yields a contradiction for the sign of some element of $\hat{\boldsymbol{F}}_p$ or $\hat{\boldsymbol{F}}_{pq}$. On the other hand, if one of the conditions (12) is not strictly satisfied, the inequality

$$\boldsymbol{H}[\boldsymbol{g}] \leq h_k \tag{21}$$

is satisfied for some index k and some integral column $\boldsymbol{g}$ with $(g_k, \ldots, g_n) = 1$, while $\pm\boldsymbol{g}$ does not coincide with the k-th column of the identity matrix. Let $\kappa_{p-1} < k \leq \kappa_p$. If $\boldsymbol{e}_p$ denotes the column that is formed by the j_p elements x_l $(\kappa_{p-1} < l \leq \kappa_p; l \neq k))$, $x_k = 1$, then (21) turns into

$$\sum_{m=1}^{r} \omega_m \boldsymbol{g}_m' \left(\boldsymbol{F}_m \boldsymbol{g}_m + 2 \sum_{q=m+1}^{r} \boldsymbol{F}_{mq} \boldsymbol{g}_q \right) \leq \omega_p \boldsymbol{F}_p[\boldsymbol{e}_p] \tag{22}$$

by using (5) and (8). In the limit we have

$$\omega_m = \mathrm{o}(\omega_{m+1}) \qquad (m = 1, \ldots, r-1)$$

and

$$\boldsymbol{F}_m \to \hat{\boldsymbol{F}}_m > 0 \qquad (m = 1, \ldots, r),$$
$$\boldsymbol{F}_{mq} \to \hat{\boldsymbol{F}}_{mq} \qquad (1 \leq m < q \leq r).$$

From (22) it follows recursively that $\boldsymbol{g}_q = 0$ for $q = r, r-1, \ldots, p+1$, hence also $(g_k, \ldots, g_{\kappa_p}) = 1$, and then

$$\hat{\boldsymbol{F}}_p[\boldsymbol{g}_p] \leq \hat{\boldsymbol{F}}_p[\boldsymbol{e}_p].$$

Since $\hat{\boldsymbol{F}}_p$ is strictly reduced it follows that $\boldsymbol{g}_p = \pm\boldsymbol{e}_p$. However, $\pm\boldsymbol{g}$ is not the k-th column of the identity matrix and hence not all $\boldsymbol{g}_m = 0\,(1 \leq m < p)$. If $\boldsymbol{g}_m = 0$ for $w < m < p$ and $\boldsymbol{g}_w \neq 0$ it follows from (22)

$$\sum_{m=1}^{w-1} \omega_m \boldsymbol{g}_m' \left(\boldsymbol{F}_m \boldsymbol{g}_m + 2 \sum_{q=m+1}^{p} \boldsymbol{F}_{mq} \boldsymbol{g}_q \right) + \omega_w \boldsymbol{g}_w' \left(\boldsymbol{F}_w \boldsymbol{g}_w \pm 2\boldsymbol{F}_{wp} \boldsymbol{e}_p \right) \leq 0,$$

hence in the limit

$$\hat{\boldsymbol{F}}_w[\pm\boldsymbol{g}_w] + 2(\pm\boldsymbol{g}_w')\hat{\boldsymbol{F}}_{wp}\boldsymbol{e}_p \leq 0.$$

This is a contradiction since the columns of $\hat{\boldsymbol{R}}_{wp}$ are strictly reduced with respect to $\hat{\boldsymbol{R}}_w$. Hence the assertion from above is proven.

Let the open neighborhood U of α on $\overline{P}\{\kappa\}$ be chosen small enough such that the part of U that is contained in P lies completely in the interior of R. Now let $\boldsymbol{H}$ tend to a boundary point β in U of type $\{\lambda\}$, where $\{\lambda\}$ is contained in $\{\kappa\}$. If one introduces linear coordinates for the type $\{\lambda\}$ in $\boldsymbol{H}$, then it follows in the limit that β is reduced. Because of continuity all boundary points of the same type that are sufficiently close to β are also reduced, and hence, β is strictly reduced. This proves that all points of a sufficiently small neighborhood of α on $\overline{P}\{\kappa\}$ are strictly reduced.

Finally, let α be a boundary point for which all conditions of the system Θ are fulfilled, but not necessarily strictly. Since these conditions are linear in the elements of $\hat{\boldsymbol{F}}_p$ and $\hat{\boldsymbol{F}}_{pq}$, it follows easily that in each neighborhood of α there are boundary points γ of the same type in which all the conditions are satisfied strictly. Since such γ are already identified as boundary points of R, this holds true for α itself. The boundary points of type $\{\kappa\}$ that are obtained by taking the closure of R in $\overline{P}\{\kappa\}$ are therefore exactly the reduced boundary points, and these are described by finitely many inequalities of the system $\Theta\{\kappa\}$. Notice that the reduction conditions for the boundary points have been explicitly formulated only in the coordinates of the respective type. All reduced boundary points of the different types together with R comprise the compactified domain $\overline{R}$ in $\overline{P}$.

6 Equivalent boundary points

Two boundary points α and β are called *equivalent* if there is a sequence of matrices $\boldsymbol{H}$ in P and equivalent matrices $\boldsymbol{H}[\boldsymbol{U}]$, such that $\boldsymbol{H} \Rightarrow \alpha$ and $\boldsymbol{H}[\boldsymbol{U}] \Rightarrow \beta$. Here, the unimodular matrix $\boldsymbol{U}$ does not need to be constant on the sequence. However, we will show that for fixed α and β the matrix $\boldsymbol{U}$ is necessarily bounded. Furthermore, it will arise that α and β are of the same type. In our analysis it is not necessary to exclude the empty type for the points α and β. However, we may assume that, say, α has non-empty type.

We apply the Jacobi transformation to $\boldsymbol{H}$ such that (2) is satisfied. For $\boldsymbol{H} \in \overline{P}(\rho)$ the following inequalities shall be proven for all real $\boldsymbol{x}$:

$$\mu \boldsymbol{T}[\boldsymbol{x}] \leq \boldsymbol{H}[\boldsymbol{x}] \leq \lambda \boldsymbol{T}[\boldsymbol{x}], \tag{23}$$

where λ and μ denote positive values that depend only on ρ and n. Hence, the assertion (23) is equivalent to the condition that all roots of the equation $|x\boldsymbol{T} - \boldsymbol{H}| = 0$ are contained in the interval $\mu \leq x \leq \lambda$. If $\boldsymbol{T}^{\frac{1}{2}} = \boldsymbol{W}$ denotes the diagonal matrix with positive diagonal elements $\sqrt{t_k}$ $(k = 1, \ldots, n)$, then for $\boldsymbol{W}\boldsymbol{D}\boldsymbol{W}^{-1} = \boldsymbol{B}$ the following formulas hold:

$$|x\boldsymbol{T} - \boldsymbol{H}| = |\boldsymbol{T}||x\boldsymbol{E} - \boldsymbol{B}'\boldsymbol{B}|, \quad |\boldsymbol{B}'\boldsymbol{B}| = |\boldsymbol{B}|^2 = |\boldsymbol{D}|^2 = 1.$$

Since the triangular matrix $\boldsymbol{B}$ arises from $\boldsymbol{D}$ by replacing the element d_{kl} by $d_{kl}\sqrt{t_k}\sqrt{t_l}^{-1}$ $(1 \leq k < l \leq n)$ and

$$t_k t_l^{-1} = u_k u_{k+1} \cdots u_{l-1},$$

the coefficients of the polynomial $|x\boldsymbol{E} - \boldsymbol{B}'\boldsymbol{B}|$ are, because of (3), bounded by constants that depend only on ρ and n. Since furthermore all roots are positive and their product equals 1, claim (23) follows. In particular, we have

$$\mu t_l \leq h_l \leq \lambda t_l \quad (l = 1, \ldots, n). \tag{24}$$

For a later purpose, we generalize the question regarding equivalence, i.e. we assume $\boldsymbol{H} \Rightarrow \alpha$ and $\boldsymbol{H}[\boldsymbol{V}] \Rightarrow \beta$ for matrices $\boldsymbol{V} = (v_{kl})$ with integer coefficients v_{kl} and given determinant $d \neq 0$. Here $|\boldsymbol{H}| = h$ and $|\boldsymbol{H}[\boldsymbol{V}]| = hd^2$. If one applies the Jacobi transformation to $\boldsymbol{H}[\boldsymbol{V}]$, one obtains

$$\boldsymbol{H}[\boldsymbol{V}] = \boldsymbol{S}[\boldsymbol{C}] \tag{25}$$

with a diagonal matrix $\boldsymbol{S}$ whose diagonal elements are $s_1, \dots, s_n$ and a triangular matrix $\boldsymbol{C}$. Because of $|\boldsymbol{V}| \neq 0$ there is a permutation $l_1, \dots, l_n$ of the numbers $1, \dots, n$ such that the product $v_{1l_1} v_{2l_2} \cdots v_{nl_n} \neq 0$. Since the v_{kl} are integer numbers, it follows that

$$v_{kl_k}^2 \geq 1 \quad (k = 1, \dots, n).$$

As the sequences $\boldsymbol{H}$ and $\boldsymbol{H}[\boldsymbol{V}]$ converge on $\overline{P}$, we have $\boldsymbol{H} \in \overline{P}(\rho)$ and $\boldsymbol{H}[\boldsymbol{V}] \in \overline{P}(\rho)$ for some fixed $\rho > 1$. If one uses (24) for $\boldsymbol{H}[\boldsymbol{V}]$ instead of $\boldsymbol{H}$ as well as (23), (25), it follows that

$$\mu \sum_{k=1}^{n} t_k v_{kl}^2 \leq \lambda s_l \quad (l = 1, \dots, n), \tag{26}$$

where the positive values λ and μ do not depend on the elements of the sequence but only on n and ρ. In particular, it follows that

$$\mu t_k \leq \lambda s_{l_k} \quad (k = 1, \dots, n). \tag{27}$$

On the other hand

$$0 < \frac{t_k}{t_{k+1}} \leq \rho, \qquad 0 < \frac{s_k}{s_{k+1}} \leq \rho \quad (k = 1, \dots, n-1). \tag{28}$$

From (27) it follows at first that

$$\mu t_k \leq \lambda s_{l_j} \frac{t_k}{t_j} \leq \lambda \rho^{j-k} s_{l_j} \quad (1 \leq k \leq j \leq n). \tag{29}$$

The minimum of the $n-k+1$ different numbers $l_k, l_{k+1}, \dots, l_n$ is not greater than k and therefore (27), (28) and (29) yield the estimate

$$t_k = \mathrm{O}(s_k) \quad (k = 1, \dots, n)$$

in the considered limit process.

On the other hand, take into account that the elements of $d\boldsymbol{V}^{-1}$ are integer numbers and

$$\boldsymbol{S}[\boldsymbol{C}][d\boldsymbol{V}^{-1}] = d^2 \boldsymbol{T}[\boldsymbol{D}].$$

If one applies the results from above to $\boldsymbol{H}[\boldsymbol{V}]$ and $d^2\boldsymbol{H}$ instead of $\boldsymbol{H}$ and $\boldsymbol{H}[\boldsymbol{V}]$, it follows that

$$s_k = \mathrm{O}(d^2 t_k) = \mathrm{O}(t_k) \quad (k = 1, \dots, n).$$

This means that s_k and t_k have the same order of magnitude in the limiting process, and in particular, the boundary points α and β are of the same type $\{\kappa\} = \{\kappa_1, \dots, \kappa_{r-1}\}$. Since the v_{kl} are integer numbers, (26) yields further

$$v_{kl} = 0 \quad (1 \leq l \leq \kappa_p < k \leq n;\ p = 1, \dots, r-1) \tag{30}$$

for almost all V in the respective sequence. If one omits finitely many elements in the sequence, one can assume (30) for all V. Hence, there is a decomposition into boxes $V = (V_{pq})$ with $V_{pq} = 0$ if $p > q$ where $V_{pq}\,(p,q = 1,\dots,r)$ consists of j_p rows and j_q columns. Since t_k and s_l have the same order of magnitude for $\kappa_{p-1} < k \le \kappa_p$ and $\kappa_{p-1} < l \le \kappa_p\,(p = 1,\dots,r)$, a further application of (26) shows that the matrices $V_{pp} = V_p\,(p = 1,\dots,r)$ remain bounded in the limiting process. If $\tilde{V}$ denotes the diagonal matrix of type $\{\kappa\}$ formed by the diagonal boxes $V_1,\dots,V_r$, we have

$$|\tilde{V}| = |V_1|\cdots|V_r| = |V| = d \neq 0.$$

We now introduce normal coordinates of type $\{\kappa\}$ for H and $H[V]$. If one forms the diagonal matrix $\hat{T}$ of type $\{\kappa\}$ as in (4) from the boxes $T_1,\dots,T_r$, and also the triangular matrix $\hat{D} = (D_{pq})$ of type $\{\kappa\}$, where $D_{pp}\,(p = 1,\dots,r)$ is an identity matrix and $D_{pq} = 0\,(p > q)$, we have

$$H = \hat{T}[\hat{D}],$$

and analogously

$$H[V] = \hat{S}[\hat{C}].$$

Then

$$\hat{T}[\tilde{V}][\tilde{V}^{-1}\hat{D}V] = \hat{S}[\hat{C}]$$

holds, where both sides appear in the normal form of type $\{\kappa\}$. From the uniqueness of this decomposition it follows that

$$\hat{T}[\tilde{V}] = \hat{S}, \quad \tilde{V}^{-1}\hat{D}V = \hat{C}, \quad V = \hat{D}^{-1}\tilde{V}\hat{C}. \tag{31}$$

Since H and $H[V]$ are convergent, the matrices $\hat{D}^{-1}$ and $\hat{C}$ remain bounded. As we already proved the boundedness of $\tilde{V}$ it now follows that V is bounded. Since V is integer, we can now pick a subsequence with fixed V.

In order to formulate the obtained result in a convenient way, we introduce a new notion. If we have a decomposition of a matrix $M = (M_{pq})$ with n rows into boxes of j_p rows and j_q columns $(p,q = 1,\dots,r)$ such that all M_{pq} below the diagonal are 0, we say *M splits of type* $\{\kappa\}$. Thus we can formulate the following theorem.

Theorem 1. *Let H converge to a boundary point of type $\{\kappa\}$ and assume that there is a sequence of integral matrices V with fixed determinant $\neq 0$ such that $H[V]$ also converges to a boundary point. Then this boundary point is also of type $\{\kappa\}$; furthermore, the elements of the sequence V lie in a finite set and almost all split of type $\{\kappa\}$.*

For a later application we remark that from the assumptions $H \in \overline{P}(\rho)$, $H[V] \in \overline{P}(\rho)$ for integral V with determinant $d \neq 0$, it follows that V is bounded with upper and lower bound only depending on n, ρ, and d.

This result will be applied for $d = \pm 1$ such that $V = U$ and all $V_p\,(p = 1,\dots,r)$ are unimodular.

If we introduce by

$$T_p = \tau_p R_p, \quad |R_p| = 1, \quad S_p = \sigma_p Q_p, \quad |Q_p| = 1 \quad (p = 1,\dots,r),$$

$$v_p = \frac{\tau_p}{\tau_{p+1}}, \quad u_p = \frac{\sigma_p}{\sigma_{p+1}} \quad (p = 1, \dots, r-1)$$

the normal coordinates of $\boldsymbol{H}$ and $\boldsymbol{H}[\boldsymbol{U}]$ in addition to the elements of $\boldsymbol{D}_{pq}$ and $\boldsymbol{C}_{pq}$ $(p < q)$, we obtain

$$\tau_p \boldsymbol{R}_p[\boldsymbol{V}_p] = \boldsymbol{T}_p[\boldsymbol{V}_p] = \boldsymbol{S}_p = \sigma_p \boldsymbol{Q}_p,$$

and hence,

$$\tau_p = \sigma_p, \quad \boldsymbol{R}_p[\boldsymbol{V}_p] = \boldsymbol{Q}_p \quad (p = 1, \dots, r),$$
$$v_p = u_p \quad (p = 1, \dots, r-1).$$

For the linear coordinates $\boldsymbol{F}_{pq}$ and $\boldsymbol{G}_{pq}$ $(1 \le p \le q \le r)$ of α and β, because of (10) and (31), we obtain in the limit (with fixed $\boldsymbol{V}$) the formula

$$\boldsymbol{V}_p \boldsymbol{Q}_p^{-1} \boldsymbol{G}_{pq} = \boldsymbol{R}_p^{-1} \sum_{m=p}^{q} \boldsymbol{F}_{pm} \boldsymbol{V}_{mq},$$

that is,

$$\boldsymbol{G}_{pq} = \boldsymbol{V}'_p \sum_{m=p}^{q} \boldsymbol{F}_{pm} \boldsymbol{V}_{mq} \quad (1 \le p \le q \le r) \tag{32}$$

with integral $\boldsymbol{V}_{mq}$ and unimodular $\boldsymbol{V}_{pp} = \boldsymbol{V}_p$. The relationship (32) can be represented more succinctly by a matrix equation: If one defines $\boldsymbol{G}_{pq} = 0$, $\boldsymbol{F}_{pq} = 0$ $(1 \le q < p \le r)$ and denotes by $\boldsymbol{G}$, $\boldsymbol{F}$ the matrices $(\boldsymbol{G}_{pq})$, $(\boldsymbol{F}_{pq})$ that split of type $\{\kappa\}$, then (32) is equivalent to

$$\boldsymbol{G} = \tilde{\boldsymbol{U}}' \boldsymbol{F} \boldsymbol{U}. \tag{33}$$

For this reason it is convenient to set directly $\alpha = \boldsymbol{F}$ and $\beta = \boldsymbol{G}$.

Conversely, we now show that two boundary points $\alpha = \boldsymbol{F}$ and $\beta = \boldsymbol{G}$ of the same type $\{\kappa\}$ are equivalent if (33) is fulfilled for some unimodular matrix $\boldsymbol{U}$ that splits of type $\{\kappa\}$. To show this we put

$$\begin{gathered} \boldsymbol{T}_p = \tau_p \boldsymbol{F}_p, \quad \boldsymbol{D}_{pq} = \boldsymbol{F}_p^{-1} \boldsymbol{F}_{pq}, \quad \boldsymbol{S}_p = \tau_p \boldsymbol{G}_p, \\ \boldsymbol{C}_{pq} = \boldsymbol{G}_p^{-1} \boldsymbol{G}_{pq} \quad (p, q = 1, \dots, r) \end{gathered} \tag{34}$$

with arbitrary positive $\tau_1, \dots, \tau_r$. Furthermore, let $\hat{\boldsymbol{T}}$, $\hat{\boldsymbol{S}}$, $\hat{\boldsymbol{F}}$, $\hat{\boldsymbol{G}}$ be the diagonal matrices of type $\{\kappa\}$ composed from the boxes $\boldsymbol{T}_p$, $\boldsymbol{S}_p$, $\boldsymbol{F}_p$, $\boldsymbol{G}_p$, and $\hat{\boldsymbol{D}} = (\boldsymbol{D}_{pq})$, $\hat{\boldsymbol{C}} = (\boldsymbol{C}_{pq})$. By (33) we then have

$$\hat{\boldsymbol{T}}[\hat{\boldsymbol{D}}\boldsymbol{U}] = \hat{\boldsymbol{T}}[\tilde{\boldsymbol{F}}^{-1} \boldsymbol{F} \boldsymbol{U}] = \hat{\boldsymbol{T}}[\tilde{\boldsymbol{U}} \tilde{\boldsymbol{G}}^{-1} \boldsymbol{G}] = \hat{\boldsymbol{S}}[\boldsymbol{C}], \tag{35}$$

and by letting $\tau_p / \tau_{p+1} \to 0$ $(p = 1, \dots, r-1)$, the assertion follows.

In the group Γ of all unimodular matrices, the matrices that split of type $\{\kappa\}$ form a subgroup Γ_κ. The mappings $\boldsymbol{H} \mapsto \boldsymbol{H}[\boldsymbol{U}]$ yield a representation of Γ_κ on P from which the representation (33) of Γ_κ is obtained as the limit on the boundary set P_κ. From the previously proven statement about boundedness it follows that this representation is discontinuous. This can be shown also directly with (33). The relation $\boldsymbol{G} = \tilde{\boldsymbol{U}}' \boldsymbol{F} \boldsymbol{U}$ for the equivalent points $\alpha = \boldsymbol{F}$ and $\beta = \boldsymbol{G}$ will be expressed by $\beta = \alpha_{\boldsymbol{U}}$. This relation holds if and only if there is a sequence $\boldsymbol{H} \Rightarrow \alpha$ such that also $\boldsymbol{H}[\boldsymbol{U}] \Rightarrow \beta$.

The matrix $\tilde{\boldsymbol{U}}$ consisting of the diagonal boxes $\boldsymbol{U}_p$ of $\boldsymbol{U}$ will be called the diagonal part of $\boldsymbol{U}$. The elements of Γ_κ with $\tilde{\boldsymbol{U}} = \boldsymbol{E}$ yield an invariant subgroup and the quotient group is obviously formed by the $\tilde{\boldsymbol{U}}$.

7 The fundamental domain of boundary points

We will now show that the reduced boundary points of type $\{\kappa\}$ on P_κ form a fundamental domain with respect to Γ_κ. Let $\alpha = \boldsymbol{F}$ be a boundary point of this type and $\boldsymbol{H} \Rightarrow \alpha$. Choose a sequence $\boldsymbol{U}$ in Γ such that $\boldsymbol{H}[\boldsymbol{U}]$ is reduced. Due to the compactness of $\overline{R}$, by choosing an appropriate subsequence, we can assume that the sequence $\boldsymbol{H}[\boldsymbol{U}]$ converges as well. Because of Theorem 1 it even follows that $\boldsymbol{H}[\boldsymbol{U}] \Rightarrow \beta$ for some reduced boundary point $\beta = \boldsymbol{G}$ of the same type, where $\boldsymbol{U}$ is constant and belongs to the subgroup Γ_κ. For the linear coordinates of α and β we have the relation (33). Therefore, for each $\alpha \in P_\kappa$ there is a Γ_κ-equivalent reduced boundary point of the same type.

Let now α already be reduced in the strict sense. Then, for $p = 1, \ldots, r$, the positive matrix $\boldsymbol{F}_p$ is reduced in the strict sense and furthermore, $\boldsymbol{G}_p$ itself is reduced. From $\boldsymbol{G}_p = \boldsymbol{F}_p[\boldsymbol{V}_p]$ it then follows that either $\boldsymbol{V}_p$ or $-\boldsymbol{V}_p$ equals the identity matrix. By replacing $\boldsymbol{U}$, if necessary, with $-\boldsymbol{U}$, we can normalize $\boldsymbol{V}_r = \boldsymbol{E}$. Now, assume that for an index p in $1, \ldots, r-1$ the following formulas are already proven:

$$\boldsymbol{V}_q = \boldsymbol{E}, \qquad \boldsymbol{V}_{mq} = 0 \qquad (q = p+1, \ldots, r;\ m = p+1, \ldots, q-1).$$

This assumption is trivially satisfied if $p = r-1$.

Because of (32) we obtain

$$\boldsymbol{V}_p \boldsymbol{G}_p^{-1} \boldsymbol{G}_{pq} = \boldsymbol{V}_{pq} + \boldsymbol{F}_p^{-1} \boldsymbol{F}_{pq} \qquad (q = p+1, \ldots, r).$$

Here the matrix $\boldsymbol{G}_p^{-1}\boldsymbol{G}_{pq} = \boldsymbol{C}_{pq}$ is reduced with respect to $\boldsymbol{G}_p$, and $\boldsymbol{F}_p^{-1}\boldsymbol{F}_{pq} = \boldsymbol{D}_{pq}$ is even strictly reduced with respect to $\boldsymbol{F}_p$; since $\boldsymbol{V}_p = \pm\boldsymbol{E}$, the first statement is also true for $\boldsymbol{V}_p\boldsymbol{C}_{pq}$. For this reason, the columns of $\boldsymbol{V}_p\boldsymbol{C}_{pq}, \boldsymbol{D}_{pq}$ are all contained in the fundamental domain $F(\boldsymbol{F}_p)$ and those of $\boldsymbol{D}_{pq}$ even in its interior. On the other hand, $\boldsymbol{V}_{pq}$ is an integral matrix and hence, $\boldsymbol{V}_{pq} = 0$. More precisely, for $q = p+1$, the first columns of $\boldsymbol{C}_{pq}, \boldsymbol{D}_{pq}$ are tightly reduced and this yields $\boldsymbol{V}_p = \boldsymbol{E}$. This concludes the induction step for decreasing values of p. One obtains $\boldsymbol{U} = \boldsymbol{E}$ and $\alpha = \beta$.

This result can also be obtained in the following way by using the result of Section 5. In any arbitrarily small neighborhood of the reduced boundary point β one can find a strictly reduced boundary point δ of the same type. If $\boldsymbol{G}^*$ denotes the matrix formed by the linear coordinates of δ and if one defines by $\boldsymbol{G}^* = \overline{\boldsymbol{U}}' \boldsymbol{F}^* \boldsymbol{U}$ the linear coordinates of a boundary point $\gamma = \boldsymbol{F}^*$, then γ is strictly reduced, too, if only δ is sufficiently close to β. If one then transfers (34) to the modified linear coordinates, the respective matrices $\hat{\boldsymbol{T}}[\hat{\boldsymbol{D}}]$ and $\hat{\boldsymbol{S}}[\hat{\boldsymbol{C}}]$ are strictly reduced for sufficiently small τ_p/τ_{p+1} $(p = 1, \ldots, r-1)$. Because of (35) we obtain $\boldsymbol{U} = \pm\boldsymbol{E}$ and $\alpha = \beta$.

Hence, the reduced boundary points of type $\{\kappa\}$ comprise a fundamental domain with respect to Γ_κ. This fundamental domain is denoted by R_κ. This R_κ is precisely the intersection of P_κ and $\overline{R}$. If one relinquishes this particular property, a fundamental domain for Γ_κ on P_κ can be found in a simpler way by using the analogous domains $F(\boldsymbol{E})$ instead of $F(\boldsymbol{G}_p)$ for the reduction of the translation group, so that F becomes a cube. This yields the following definition of a domain N_κ which we want to call a *normal fundamental domain for* Γ_κ. As before, let the $\boldsymbol{G}_p$ $(p = 1, \ldots, r)$ be reduced in the sense of Minkowski; the elements of all matrices $\boldsymbol{C}_{pq} = \boldsymbol{G}_p^{-1}\boldsymbol{G}_{pq}$ $(1 \le p < q \le r)$ are contained in the interval $-\frac{1}{2} \le x \le \frac{1}{2}$ and for the first column at $q = p+1$ the last element is even contained in the interval $0 \le x \le \frac{1}{2}$. One needs to show that N_κ indeed is a fundamental domain for Γ_κ on P_κ. Let again $\boldsymbol{F}_{pq}$ be linear coordinates of a boundary point α. Then choose a unimodular $\boldsymbol{V}_{pp} = \boldsymbol{V}_p$ $(p = 1,\dots,r)$ such that $\boldsymbol{F}_p[\boldsymbol{V}_p] = \boldsymbol{G}_p$ is reduced in the sense of Minkowski and determine $\boldsymbol{V}_{pq}$ $(q = p+1,\dots,r)$ recursively with respect to decreasing p such that the elements of the matrices

$$\boldsymbol{V}_p^{-1}\Big(\boldsymbol{V}_{pq} + \sum_{m=p+1}^{q} \boldsymbol{F}_p^{-1}\boldsymbol{F}_{pm}\boldsymbol{V}_{mq}\Big) = \boldsymbol{C}_{pq}$$

lie all between $-\frac{1}{2}$ and $\frac{1}{2}$. By replacing $\boldsymbol{V}_p$ with $-\boldsymbol{V}_p$, if necessary, one can assure that the last element of the first row of $\boldsymbol{C}_{pq}$ lies in the interval $0 \le x \le \frac{1}{2}$. Hence, (32) is satisfied and the boundary point β with linear coordinates $\boldsymbol{G}_{pq}$, which is equivalent to α, lies in N_κ. Conversely, one shows in an analogous way that two inner points of N_κ are not equivalent.

The normal fundamental domain N_κ is evidently the direct product of the r Minkowski-domains R_l $(l = j_1,\dots,j_r)$ and the intervals for the elements of the $\boldsymbol{C}_{pq}$. If one now forms the matrices $\hat{\boldsymbol{T}}[\hat{\boldsymbol{D}}] = \boldsymbol{H}$ for the points of N_κ with $\boldsymbol{T}_p = \tau_p\boldsymbol{F}_p$, $\tau_p > 0$ $(p = 1,\dots,r)$, by using our earlier notation, then one obtains a fundamental domain F_κ with respect to Γ_κ on P. This fundamental domain is thus the direct product of N_κ with the $r-1$ half-lines $v_p > 0$ $(p = 1,\dots,r-1)$ and the normal coordinates of $\boldsymbol{H}$ are just the normal coordinates of the trace point on N_κ.

8 Geodesic lines

In the definition of the Minkowski domain R, the extremal property (12) is of fundamental significance. If one also considers the properties of the geodesic lines in P, further relations are revealed.

By ignoring in the first instance the condition $|\boldsymbol{H}| = h$, we introduce a Riemannian metric on the space of positive $\boldsymbol{H}$ via the trace of the matrix $(\boldsymbol{H}^{-1}\dot{\boldsymbol{H}})^2$, where the dot denotes the differentiation with respect to some parameter. This trace is invariant under all mappings $\boldsymbol{H} \mapsto \boldsymbol{H}[\boldsymbol{C}]$ with real, invertible $\boldsymbol{C}$, and is, on the other hand, thereby uniquely determined up to some positive factor. As differential equation for the geodesic lines we obtain

$$\ddot{\boldsymbol{H}} = \boldsymbol{H}^{-1}[\dot{\boldsymbol{H}}].$$

This differential equation can be integrated easily, if one writes the equation in the form

$$(\boldsymbol{H}^{-1}\dot{\boldsymbol{H}})^{\cdot} = 0.$$

It then follows that

$$\boldsymbol{H} = \boldsymbol{T}^{\rho}[\boldsymbol{C}] = \boldsymbol{H}(\rho), \tag{36}$$

where $\boldsymbol{C}$ denotes an arbitrary real invertible matrix and $\boldsymbol{T}$ is an arbitrary positive diagonal matrix. Furthermore, the real variable ρ is the parameter of the curve and $\boldsymbol{T}^{\rho}$ denotes the diagonal matrix with the diagonal elements $t_1^{\rho},\dots,t_n^{\rho}$, such that $\boldsymbol{T}$ itself has the diagonal elements $t_1,\dots,t_n$. We can assume that these are increasingly ordered. If one puts

$$H(0) = C'C = P, \qquad H(1) = T[C] = Q,$$

then the arc length $d = d(P,Q)$ from P to Q is given by the formula

$$d^2 = \sum_{k=1}^{n} \log(t_k)^2. \tag{37}$$

If $T \neq E$, then $d > 0$. Furthermore, if one normalizes T by demanding $d = 1$, then $\rho = s$ is the arc length of the geodesic line from P to H.

Conversely, if P and Q are two distinct positive matrices one can apply a principal axis transformation so that $P = C'C$ and $Q = T[C]$. Because of $P^{-1}Q = C^{-1}TC$ the numbers $t_1 < \cdots < t_n$ are the eigenvalues of $P^{-1}Q$, i.e. the roots of the polynomial $|tP - Q|$. The geodesic line given by (36) then passes through P and Q. If the geodesic line

$$\overline{H} = \overline{T}^{\sigma}[\overline{C}] = \overline{H}(\sigma)$$

has the same property, then let

$$\overline{T}^{\sigma_1}[\overline{C}] = P, \quad \overline{T}^{\sigma_2}[\overline{C}] = Q \qquad (\sigma_2 > \sigma_1).$$

It follows that

$$\overline{T}^{\sigma_2-\sigma_1}\overline{C}C^{-1} = \overline{C}C^{-1}T,$$

hence,

$$T = \overline{T}^{\sigma_2-\sigma_1}$$

and therefore $\overline{C}C^{-1}$ and T^{ρ} commute. Furthermore,

$$H = C'T^{\rho}C = PC^{-1}T^{\rho}C = \overline{C}'\overline{T}^{\sigma_1}\overline{C}C^{-1}T^{\rho}C = \overline{C}'\overline{T}^{\sigma_1+(\sigma_2-\sigma_1)\rho}\overline{C} = \overline{H}$$

with $\sigma = \sigma_1 + (\sigma_2 - \sigma_1)\rho$. Therefore, the geodesic line through two arbitrary distinct points P and Q is uniquely determined and (37) yields the distance $d(P,Q)$. For those distances the triangle inequality holds. Furthermore

$$d(P[B], Q[B]) = d(P,Q)$$

for all real invertible B and in particular, $d(aP, aQ) = d(P,Q)$ for positive scalar factors a. In the case $|P| = |Q|$ it follows that $|T| = |P^{-1}Q| = 1$ and hence also

$$|T^{\rho}| = |T|^{\rho} = 1, \quad |H| = |T^{\rho}[C]| = |C'C| = |P|.$$

Therefore, the geodesic line through two points with $|H| = h$ is contained completely in the level surface of the determinant defined by this equation.

Let the multiplicities of the eigenvalues $t_1,\dots,t_n$ correspond to the type $\{\kappa\}$ such that

$$\begin{aligned} t_k &= t_l \qquad (\kappa_{p-1} < k < l \le \kappa_p;\ p = 1,\dots,r),\\ t_k &< t_l \qquad (k \le \kappa_p < l;\ p = 1,\dots,r-1). \end{aligned}$$

Let τ_p, $p = 1,\dots,r$, be positive variables that will be used as the diagonal elements $s_1,\dots,s_n$ with multiplicities $j_p = \kappa_p - \kappa_{p-1}$ of a diagonal matrix $\boldsymbol{W}$, i.e.

$$s_k = \tau_p \qquad (\kappa_{p-1} < k \le \kappa_p;\ p = 1,\dots,r).$$

The matrices $\boldsymbol{G} = \boldsymbol{W}[\boldsymbol{C}]$ can also be characterized as the positive linear combinations of $\boldsymbol{Q}$, $\boldsymbol{Q}\boldsymbol{P}^{-1}\boldsymbol{Q}$, $\boldsymbol{Q}\boldsymbol{P}^{-1}\boldsymbol{Q}\boldsymbol{P}^{-1}\boldsymbol{Q}$,... with real scalar coefficients. The set of these matrices is denoted by $L = L(\boldsymbol{P},\boldsymbol{Q})$. Then the geodesic line connecting two points $\boldsymbol{G}_1 = \boldsymbol{W}_1[\boldsymbol{C}]$ and $\boldsymbol{G}_2 = \boldsymbol{W}_2[\boldsymbol{C}]$ in L is given by

$$\boldsymbol{G} = (\boldsymbol{W}_1^{1-\rho}\boldsymbol{W}_2^{\rho})[\boldsymbol{C}] = \boldsymbol{G}(\rho)$$

and is completely contained in L. In particular, this holds for $\boldsymbol{H}(\rho)$ as $\boldsymbol{P}$ and $\boldsymbol{Q}$ are contained in L. In the following, we still consider only points in $|\boldsymbol{H}| = h$ and require $|\boldsymbol{P}| = |\boldsymbol{Q}| = |\boldsymbol{G}| = h$, i.e.

$$|\boldsymbol{C}|^2 = h, \qquad t_1\cdots t_n = s_1\cdots s_n = 1.$$

The intersection of L and P is of dimension $r-1$ and will be called the *geodesic plane* $G = G(\boldsymbol{P},\boldsymbol{Q})$ through $\boldsymbol{P}$ and $\boldsymbol{Q}$. It is easy to see that L as well as G are equipped with the Euclidean metric.

Now let α be a boundary point of type $\{\kappa\}$ with linear coordinates $\boldsymbol{F}_{pq}$. Once more we put

$$\boldsymbol{D}_{pq} = \boldsymbol{F}_p^{-1}\boldsymbol{F}_{pq}, \quad \boldsymbol{T}_p = \tau_p\boldsymbol{F}_p, \quad \hat{\boldsymbol{T}}[\hat{\boldsymbol{D}}] = \boldsymbol{H},$$

so that in the limit

$$v_p = \tau_p/\tau_{p+1} \to 0 \quad (p = 1,\dots,r-1)$$

the point $\boldsymbol{H}$ converges to α. Now let

$$\tau_p = b_p\theta_p^s, \quad b_p > 0 \quad (p = 1,\dots,r), \quad 0 < \theta_1 < \theta_2 < \cdots < \theta_r,$$

$$b_1^{j_1}\cdots b_r^{j_r} = h, \quad \theta_1^{j_1}\cdots\theta_r^{j_r} = 1, \quad \sum_{p=1}^{r} j_p\log(\theta_p)^2 = 1.$$

For every system of fixed b_p, θ_p, $\boldsymbol{H} = \boldsymbol{H}(s)$ is a geodesic line with arc length s that tends to α for $s \to \infty$. All of these geodesic lines fill out the geodesic plane $G(\boldsymbol{H}(0),\boldsymbol{H}(s))$ which only depends on α. These geodesic planes consist precisely of all points $\boldsymbol{H}$ with arbitrary v_p $(p = 1,\dots,r-1)$ whose normal coordinates are given by those of α. They will be denoted by $G(\alpha)$ for short. So the fundamental domain F_κ introduced in the previous section is fibered by these $G(\alpha)$.

Two boundary points α and β can be assigned a distance in a straightforward manner. For two neighborhoods A and B of α and β in $\overline{P}$ define the distance $d(A,B)$ as the lower bound of distances $d(\boldsymbol{P},\boldsymbol{Q})$ with $\boldsymbol{P} \in A$, $\boldsymbol{Q} \in B$. Then let $d(\alpha,\beta)$ be the upper bound of these $d(A,B)$ for the set of all such pairs A, B. To compute $d(\alpha,\beta)$, we use the Jacobi transformation $\boldsymbol{P} = \boldsymbol{T}[\boldsymbol{D}]$, $\boldsymbol{Q} = \overline{\boldsymbol{T}}[\overline{\boldsymbol{D}}]$. Let t_k, $\overline{t}_k$ $(k = 1,\dots,n)$ be the diagonal elements of the diagonal matrices $\boldsymbol{T}$, $\overline{\boldsymbol{T}}$, and define a triangular matrix $\boldsymbol{D}_0 = \overline{\boldsymbol{D}}\boldsymbol{D}^{-1}$. If $d(\alpha,\beta)$ is finite, then there must be sequences $\boldsymbol{P} \Rightarrow \alpha$, $\boldsymbol{Q} \Rightarrow \beta$, on which the matrices $\overline{\boldsymbol{T}}[\boldsymbol{D}_0\boldsymbol{T}^{-\frac{1}{2}}]$ remain bounded, and in particular the values $\overline{t}_k t_k^{-1}$ and $t_k\overline{t}_k^{-1}$ $(k = 1,\dots,n)$ are bounded, so that they are of the same order when taking the limit. It follows that

α and β are of the same type $\{\kappa\}$. Now introduce normal coordinates $\boldsymbol{R}_p$, $\boldsymbol{D}_{pq}$ and $\overline{\boldsymbol{R}}_p$, $\overline{\boldsymbol{D}}_{pq}$ of type $\{\kappa\}$ for α and β, and use the Jacobi transformation $\boldsymbol{R}_p = \boldsymbol{S}_p[\boldsymbol{C}_p]$. The accumulation points of the matrices $\overline{\boldsymbol{T}}[\boldsymbol{D}_0\boldsymbol{T}^{-\frac{1}{2}}]$ for arbitrary limiting processes are precisely the diagonal matrices that are composed of diagonal boxes $x_p\overline{\boldsymbol{R}}_p[\boldsymbol{C}_p^{-1}\boldsymbol{S}_p^{-\frac{1}{2}}]$ with arbitrary positive numbers $x_p^{j_p}$ of product 1. The eigenvalues of these boxes are those of the $x_p\overline{\boldsymbol{R}}_p\boldsymbol{R}_p^{-1}$, and since $|\boldsymbol{R}_p| = |\overline{\boldsymbol{R}}_p| = 1$,

$$d(x_p\overline{\boldsymbol{R}}_p,\boldsymbol{R}_p)^2 = j_p\log(x_p)^2 + d(\overline{\boldsymbol{R}}_p,\boldsymbol{R}_p)^2 \geq d(\overline{\boldsymbol{R}}_p,\boldsymbol{R}_p)^2,$$

where equality holds only for $x_p = 1$. From this it readily follows that

$$d(\alpha,\beta)^2 = \sum_{p=1}^{r} d(\boldsymbol{R}_p,\overline{\boldsymbol{R}}_p)^2,$$

so that two boundary points of the same type $\{\kappa\}$ have distance 0 if and only if their normal coordinates are $\boldsymbol{R}_p = \overline{\boldsymbol{R}}_p$ $(p = 1,\ldots,r)$.

9 Parabolic transformations

Let $\boldsymbol{W}$ be unimodular. We shall investigate under which conditions on $\boldsymbol{W}$ there exists a sequence $\boldsymbol{H}$ in P for which the distance $d(\boldsymbol{H},\boldsymbol{H}[\boldsymbol{W}])$ tends to 0. For later purposes we first consider a more general question that is answered by the following theorem:

Theorem 2. *Let $\boldsymbol{P}$ and $\boldsymbol{Q}$ run through two sequences of P that converge to boundary points α and β, and assume there exists a sequence of unimodular $\boldsymbol{U}$ such that the corresponding distances $d(\boldsymbol{P},\boldsymbol{Q}[\boldsymbol{U}])$ remain bounded. Then α and β have the same type $\{\kappa\}$, and almost all $\boldsymbol{U}$ belong to Γ_κ with bounded diagonal part $\tilde{\boldsymbol{U}}$. Moreover, $d(\boldsymbol{P},\boldsymbol{Q})$ remains bounded. If the distance $d(\boldsymbol{P},\boldsymbol{Q}[\boldsymbol{U}])$ converges to 0, then the same holds for $d(\boldsymbol{P},\boldsymbol{Q}[\tilde{\boldsymbol{U}}])$.*

For the proof we use the Jacobi transformation for $\boldsymbol{P}$ and $\boldsymbol{Q}$,

$$\boldsymbol{P} = \boldsymbol{T}[\boldsymbol{D}], \quad \boldsymbol{Q} = \boldsymbol{S}[\boldsymbol{M}]$$

with triangular matrices $\boldsymbol{D} = (d_{kl})$, $\boldsymbol{M} = (m_{kl})$ and diagonal matrices $\boldsymbol{T}$, $\boldsymbol{S}$ whose diagonal elements are t_k, s_k. Moreover, let $\{\kappa\} = \{\kappa_1,\ldots,\kappa_{r-1}\}$ and $\{\lambda\} = \{\lambda_1,\ldots,\lambda_{v-1}\}$ be the types of α and β, where we also admit the empty type. By assumption, the normal coordinates $d_{kl}, m_{kl}, t_k/t_{k+1}, s_k/s_{k+1}$ $(1 \leq k < l \leq n)$ converge to certain limits. Here, the limits of t_k/t_{k+1} and s_l/s_{l+1} are zero for $k = \kappa_1,\ldots,\kappa_{r-1}$, $l = \lambda_1,\ldots,\lambda_{v-1}$, and positive else. The numbers κ_{r-1} and λ_{v-1} are both less than n.

By our assumption on the distance

$$d(\boldsymbol{P},\boldsymbol{Q}[\boldsymbol{U}]) = d(\boldsymbol{E},\boldsymbol{S}[\boldsymbol{M}\boldsymbol{U}\boldsymbol{D}^{-1}\boldsymbol{T}^{-\frac{1}{2}}]),$$

the matrices $\boldsymbol{S}^{\frac{1}{2}}\boldsymbol{M}\boldsymbol{U}\boldsymbol{D}^{-1}\boldsymbol{T}^{-\frac{1}{2}} = \boldsymbol{L}$ and $\boldsymbol{L}^{-1}$ remain bounded when taking the limit. In particular, if we put $\boldsymbol{U} = (u_{kl})$, $\boldsymbol{D}^{-1} = (c_{kl})$, then

$$\sum_{\mu=k}^{n}\sum_{\nu=1}^{l} m_{k\mu}u_{\mu\nu}c_{\nu l} = \mathrm{O}\left(t_l^{\frac{1}{2}} s_k^{-\frac{1}{2}}\right) \quad (k, l = 1, \ldots, n). \tag{38}$$

We now show that the t_k and s_l are of the same order for $k = \kappa_{r-1}+1, \ldots, n$ and $l = \lambda_{\upsilon-1}+1, \ldots, n$. If this was not the case, then, after exchanging $\boldsymbol{P}$ and $\boldsymbol{Q}$ if necessary, for some subsequence the ratio t_n/s_n would converge to 0. If we then apply (38) with $k = n$ and $l = 1, \ldots, n$, it follows that the last row $\boldsymbol{u}'$ of $\boldsymbol{U}$ satisfies $\boldsymbol{u}'\boldsymbol{D}^{-1} \to 0$. But since $\boldsymbol{D}$ is bounded and $\boldsymbol{u}$ integral, it follows that $\boldsymbol{u} = 0$ for almost all elements of the sequence and thus the contradiction $|\boldsymbol{U}| = 0$. Hence

$$t_n = \mathrm{O}(s_n), \quad s_n = \mathrm{O}(t_n)$$

and

$$t_k = \mathrm{o}(s_l) \quad (k \le \kappa_{r-1},\ l > \lambda_{\upsilon-1}). \tag{39}$$

In an analogous manner we find $\kappa_{r-1} = \lambda_{\upsilon-1}$. For if $\boldsymbol{M}_2 = (m_{k\mu})$, $\boldsymbol{U}_{21} = (u_{\mu\nu})$, $\boldsymbol{C}_1 = (c_{\nu l})$ are the matrices formed with $k, \mu = \lambda_{\upsilon-1}+1, \ldots, n$ and $\nu, l = 1, \ldots, \kappa_{r-1}$, then we obtain from (38) and (39)

$$\boldsymbol{M}_2\boldsymbol{U}_{21}\boldsymbol{C}_1 \to 0.$$

Since $\boldsymbol{M}_2$ and $\boldsymbol{C}_1$ are bounded triangular matrices here, it follows that $\boldsymbol{U}_{21} \to 0$, that is, $\boldsymbol{U}_{21} = 0$ for almost all elements. Due to the linear dependence of the last $n - \lambda_{\upsilon-1}$ rows of $\boldsymbol{U}$, it follows that $n - \lambda_{\upsilon-1} \le n - \kappa_{r-1}$, that is, $\kappa_{r-1} \le \lambda_{\upsilon-1}$. By exchanging $\boldsymbol{P}$ and $\boldsymbol{Q}$, it also follows that $\lambda_{\upsilon-1} \le \kappa_{r-1}$, and hence the claim. Moreover, we have found that almost all elements of the sequence have a decomposition

$$\boldsymbol{U} = \begin{pmatrix} \boldsymbol{U}_1 & * \\ 0 & \boldsymbol{U}_2 \end{pmatrix}$$

with unimodular $\boldsymbol{U}_1, \boldsymbol{U}_2$, of which $\boldsymbol{U}_2$ consists of $n - \kappa_{r-1} = j_r$ rows. After applying (38) once more, we find that $\boldsymbol{U}_2$ is bounded. Now consider the submatrix consisting of the first κ_{r-1} rows of $\boldsymbol{L}$ and use induction on n. In this way we find that the types $\{\kappa\}$ and $\{\lambda\}$ are equal, and moreover that almost all $\boldsymbol{U}$ belong to Γ_κ and have a bounded diagonal part $\tilde{\boldsymbol{U}}$. Furthermore, for all $p = 1, \ldots, r$, the t_k and s_l (with $\kappa_{p-1} < k \le \kappa_p$; $\kappa_{p-1} < l \le \kappa_p$) are of the same order, from the boundedness of

$$d(\boldsymbol{P}, \boldsymbol{Q}) = d(\boldsymbol{E}, \boldsymbol{S}[\boldsymbol{M}\boldsymbol{D}^{-1}\boldsymbol{T}^{-\frac{1}{2}}])$$

is evident.

If the distance $d(\boldsymbol{P}, \boldsymbol{Q}[\boldsymbol{U}])$ now tends to 0 for the sequence under consideration, then

$$\boldsymbol{L}'\boldsymbol{L} = \boldsymbol{S}[\boldsymbol{M}\boldsymbol{U}\boldsymbol{D}^{-1}\boldsymbol{T}^{-\frac{1}{2}}] \to \boldsymbol{E}.$$

If we decompose $\boldsymbol{L} = (\boldsymbol{W}_{pq})$, where the boxes $\boldsymbol{W}_{pq}$ $(p, q = 1, \ldots, r)$ consist of j_p rows and j_q columns and are 0 for $p > q$, then it follows that

$$\begin{aligned} &\sum_{m=1}^{p} \boldsymbol{W}'_{mp}\boldsymbol{W}_{mp} \to \boldsymbol{E}_p \quad (p = 1, \ldots, r), \\ &\sum_{m=1}^{p} \boldsymbol{W}'_{mp}\boldsymbol{W}_{mq} \to 0 \quad (1 \le p < q \le r) \end{aligned} \tag{40}$$

with the identity matrix $\boldsymbol{E}_p$ of j_p rows. If the relation

$$\boldsymbol{W}'_{kk}\boldsymbol{W}_{kk} \to \boldsymbol{E}_k, \quad \boldsymbol{W}_{kl} \to 0 \quad (l = k+1,\ldots,r) \tag{41}$$

has been proven for $k = 1,\ldots,p-1$, then it follows from (40) that (41) holds for $k = p$. This implies

$$\boldsymbol{S}^{\frac{1}{2}}\boldsymbol{M}\tilde{\boldsymbol{U}}\boldsymbol{D}^{-1}\boldsymbol{T}^{-\frac{1}{2}} - \boldsymbol{L} \to 0, \quad \boldsymbol{S}[\boldsymbol{M}\tilde{\boldsymbol{U}}\boldsymbol{D}^{-1}\boldsymbol{T}^{-\frac{1}{2}}] \to \boldsymbol{E}$$

and

$$d(\boldsymbol{P},\boldsymbol{Q}[\tilde{\boldsymbol{U}}]) = d(\boldsymbol{E},\boldsymbol{S}[\boldsymbol{M}\tilde{\boldsymbol{U}}\boldsymbol{D}^{-1}\boldsymbol{T}^{-\frac{1}{2}}]) \to 0.$$

So all claims of the theorem are proved. Observe that the corresponding statements are correct for the empty type.

We now consider the problem that was posed at the beginning of this section. Let $\boldsymbol{W}$ be a fixed unimodular matrix and assume $\boldsymbol{H}$ runs through a sequence in P for which the distance $d(\boldsymbol{H},\boldsymbol{H}[\boldsymbol{W}])$ converges to 0. Let a unimodular matrix $\boldsymbol{V}$ be given for every element of the sequence, such that $\boldsymbol{H}[\boldsymbol{V}]$ lies in the Minkowski domain R. We choose a subsequence in which $\boldsymbol{H}[\boldsymbol{V}]$ converges to a boundary point α of type $\{\kappa\}$, and apply Theorem 2 with $\boldsymbol{P} = \boldsymbol{Q} = \boldsymbol{H}[\boldsymbol{V}]$, $\boldsymbol{U} = \boldsymbol{V}^{-1}\boldsymbol{W}\boldsymbol{V}$. Then $\beta = \alpha$, and

$$d(\boldsymbol{P},\boldsymbol{Q}[\boldsymbol{U}]) = d(\boldsymbol{P}[\boldsymbol{V}^{-1}],\boldsymbol{P}[\boldsymbol{U}\boldsymbol{V}^{-1}]) = d(\boldsymbol{H},\boldsymbol{H}[\boldsymbol{W}]) \to 0.$$

So there exists a subsequence with $\boldsymbol{U} \in \Gamma_\kappa$ and constant diagonal part, such that

$$d(\boldsymbol{P},\boldsymbol{P}[\tilde{\boldsymbol{U}}]) = d(\boldsymbol{P},\boldsymbol{Q}[\tilde{\boldsymbol{U}}]) \to 0. \tag{42}$$

Introduce normal coordinates of type $\{\kappa\}$ for $\boldsymbol{P}$, which gives rise to the decomposition

$$\boldsymbol{P} = \hat{\boldsymbol{T}}[\hat{\boldsymbol{D}}], \quad \boldsymbol{T}_p = \tau_p \boldsymbol{R}_p, \quad |\boldsymbol{R}_p| = 1 \quad (p = 1,\ldots,r).$$

Here, the matrices $\hat{\boldsymbol{D}}$, $\boldsymbol{R}_p$ converge, and the $r-1$ ratios τ_p/τ_{p+1} $(p = 1,\ldots,r-1)$ converge to 0. By applying the same reasoning as for (41), (42) leads to the statement

$$\hat{\boldsymbol{T}}[\tilde{\boldsymbol{U}}]\hat{\boldsymbol{T}}^{-1} \to \boldsymbol{E}, \quad \boldsymbol{R}_p[\boldsymbol{U}_p] - \boldsymbol{R}_p \to 0.$$

If $\boldsymbol{R}_p \to \boldsymbol{F}_p$, then $\boldsymbol{F}_p[\boldsymbol{U}_p] = \boldsymbol{F}_p > 0$, and from this follows the boundedness of all powers $\boldsymbol{U}_p^m$ $(m = 0,\pm1,\pm2,\ldots)$. Thus there exists an integer m such that $\tilde{\boldsymbol{U}}^m = \boldsymbol{E}$. So the eigenvalues of $\tilde{\boldsymbol{U}}$ are the m-th roots of unity and the elementary divisors are all linear. Because of

$$\lambda\boldsymbol{E} - \boldsymbol{W} = \boldsymbol{V}(\lambda\boldsymbol{E} - \boldsymbol{U})\boldsymbol{V}^{-1}, \quad |\lambda\boldsymbol{E} - \boldsymbol{U}| = |\lambda\boldsymbol{E} - \tilde{\boldsymbol{U}}|,$$

the eigenvalues of $\boldsymbol{W}$ all lie on the unit circle. In particular, if $\{\kappa\}$ is the empty type, then $\tilde{\boldsymbol{U}} = \boldsymbol{U}$ and hence $\boldsymbol{W}^m = \boldsymbol{E}$. The periodicity of $\tilde{\boldsymbol{U}}$ can also be derived in a shorter way from (42), since by virtue of the triangular inequality, for every fixed $l = 1,2,\ldots$ the relationship

$$d(\boldsymbol{P},\boldsymbol{P}[\tilde{\boldsymbol{U}}^l]) \le \sum_{k=1}^{l} d(\boldsymbol{P}[\tilde{\boldsymbol{U}}^{k-1}],\boldsymbol{P}[\tilde{\boldsymbol{U}}^k]) = l\,d(\boldsymbol{P},\boldsymbol{P}[\tilde{\boldsymbol{U}}]) \to 0$$

holds, and thus the boundedness of $\tilde{U}^l$ follows from Theorem 2.

Conversely, assume now that the eigenvalues of the unimodular matrix $\boldsymbol{W}$ all lie on the unit circle. As the roots of $|\lambda \boldsymbol{E} - \boldsymbol{W}| = 0$, the eigenvalues are algebraic integers whose conjugates have absolute value 1. Thus they are roots of unity by a well-known theorem of Kronecker. Hence there exists an integer m such that the eigenvalues of the power $\boldsymbol{W}^m$ are all 1. If now $\boldsymbol{W}^m = \boldsymbol{E}$, put

$$\sum_{k=1}^{m} c\boldsymbol{E}[\boldsymbol{W}^k] = \boldsymbol{H}, \tag{43}$$

where the number c is chosen such that $|\boldsymbol{H}| = h$. This is possible since the left hand side of (43) is positive, and then also $\boldsymbol{H} > 0$. Clearly, $\boldsymbol{H}$ is then a fixed point for $\boldsymbol{W}$, namely $\boldsymbol{H}[\boldsymbol{W}] = \boldsymbol{H}$ and $d(\boldsymbol{H}, \boldsymbol{H}[\boldsymbol{W}]) = 0$. We call $\boldsymbol{W}$ *elliptic*.

Now we turn to the more interesting case in which $\boldsymbol{W}^m \neq \boldsymbol{E}$, while all eigenvalues of $\boldsymbol{W}^m$ are equal to 1. Then $\boldsymbol{W}$ shall be called *parabolic*. In this case, the elementary divisors of $\lambda \boldsymbol{E} - \boldsymbol{W}$ are not all linear. In particular, the polynomial $|\lambda \boldsymbol{E} - \boldsymbol{W}|$ contains a multiple factor. By results from the theory of elementary divisors we can find an invertible matrix $\boldsymbol{K}$ with rational elements, such that there is a decomposition

$$\boldsymbol{K}^{-1}\boldsymbol{W}\boldsymbol{K} = \boldsymbol{U} = (\boldsymbol{U}_{pq}), \quad \boldsymbol{U}_{pp} = \boldsymbol{U}_p,$$

where p, q run through 1 to r and $\boldsymbol{U}_{pq} = 0$ for $p > q$. Moreover, the r polynomials $|\lambda \boldsymbol{E}_p - \boldsymbol{U}_p|$ are all irreducible over the field of rational numbers, so that the elementary divisors of $\lambda \boldsymbol{E}_p - \boldsymbol{U}_p$ are linear, and hence $\boldsymbol{U}_p^m = \boldsymbol{E}_p$ holds. On the other hand, there exists a unimodular matrix $\boldsymbol{V}$ such that $\boldsymbol{V}^{-1}\boldsymbol{K} = \boldsymbol{C}$ becomes a triangular matrix in which all elements below the diagonal vanish. If we replace $\boldsymbol{K}$ by $\boldsymbol{K}\boldsymbol{C}^{-1} = \boldsymbol{V}$, then the matrix $\boldsymbol{V}^{-1}\boldsymbol{W}\boldsymbol{V} = \boldsymbol{C}\boldsymbol{U}\boldsymbol{C}^{-1}$ decomposes in the same manner as $\boldsymbol{U}$ itself. So we may assume that already $\boldsymbol{K} = \boldsymbol{V}$ and $\boldsymbol{C} = \boldsymbol{E}$, so that $\boldsymbol{U} = \boldsymbol{V}^{-1}\boldsymbol{W}\boldsymbol{V}$ is also unimodular. For the diagonal part $\tilde{\boldsymbol{U}}$ of $\boldsymbol{U}$ formed by the $\boldsymbol{U}_p$ it holds that $\tilde{\boldsymbol{U}}^m = \boldsymbol{E}$, whereas $\boldsymbol{U}^m \neq \boldsymbol{E}$. Now let $\boldsymbol{F}_p$ be a fixed point for $\boldsymbol{U}_p$ with $\boldsymbol{F}_p > 0$ and $|\boldsymbol{F}_p| = 1$. If we let $\boldsymbol{T}_p = \tau_p \boldsymbol{F}_p$ $(p = 1, \ldots, r)$ and define $\hat{\boldsymbol{T}}$ as we did before, then it also holds for the matrix $\boldsymbol{H} = \hat{\boldsymbol{T}}[\boldsymbol{V}^{-1}]$ that when taking the limit $\tau_p/\tau_{p+1} \to 0$ $(p = 1, \ldots, r-1)$, we obtain the desired relation

$$d(\boldsymbol{H}, \boldsymbol{H}[\boldsymbol{W}]) = d(\hat{\boldsymbol{T}}, \hat{\boldsymbol{T}}[\boldsymbol{U}]) \to 0.$$

This proves the following result.

Theorem 3. *For a unimodular matrix $\boldsymbol{U}$ there exists a sequence of positive $\boldsymbol{H}$ with $d(\boldsymbol{H}, \boldsymbol{H}[\boldsymbol{U}]) \to 0$ if and only if $\boldsymbol{U}$ is elliptic or parabolic.*

10 Reduced distances

By the *reduced distance* of two points $\boldsymbol{P}$ and $\boldsymbol{Q}$ of P we mean the minimum of the distances $d(\boldsymbol{P}, \boldsymbol{Q}[\boldsymbol{U}])$, where $\boldsymbol{U}$ runs through all unimodular matrices. The reduced distance is denoted by $f(\boldsymbol{P}, \boldsymbol{Q})$. This can be viewed as the distance from $\boldsymbol{P}$ to the space P/Γ, since clearly $f(\boldsymbol{P}, \boldsymbol{Q}) = f(\boldsymbol{P}, \boldsymbol{Q}[\boldsymbol{U}])$.

Theorem 4. *Fix a point* $\boldsymbol{P}$ *in* P *and let the point* $\boldsymbol{R}$ *vary in* R. *Then the difference of the distance* $d(\boldsymbol{P},\boldsymbol{R})$ *and the reduced distance* $f(\boldsymbol{P},\boldsymbol{R})$ *is smaller than a bound independent of* $\boldsymbol{R}$.

By the triangle inequality it is enough to prove the theorem for any particular point $\boldsymbol{P}$. So let $\boldsymbol{P}=\boldsymbol{E}$, and write $d(\boldsymbol{E},\boldsymbol{R})=d(\boldsymbol{R})$ for short. We need to prove that the difference $d(\boldsymbol{R}[\boldsymbol{U}])-d(\boldsymbol{R})$ is above some bound that does not depend on $\boldsymbol{R}$ or $\boldsymbol{U}$, where $\boldsymbol{U}\in\Gamma$ and $\boldsymbol{R}\in R$.

By means of the Jacobi transformation we get $\boldsymbol{R}=\boldsymbol{T}[\boldsymbol{D}]$, where the normal coordinates d_{kl} and $u_k = t_k/t_{k+1}$ $(1\le k<l\le n)$ are bounded. Assume the eigenvalues $\tau_1,\dots,\tau_n$ of $\boldsymbol{R}$ are sorted in ascending order, and let $r_1,\dots,r_n$ be their elementary symmetric polynomials. To conveniently formulate the Laplace expansion, we will for every $k=1,\dots,n$ let $D_{\overline{p}\overline{q}}$ denote the k-row subdeterminant of $\boldsymbol{D}$, where $\overline{p}$ and $\overline{q}$ are abbreviations for the sequences of indices $p_1,\dots,p_k$ and $q_1,\dots,q_k$ of those rows and columns of $\boldsymbol{D}$ that form the respective subdeterminant. Here, $p_1<p_2<\dots<p_k$ and $q_1<q_2<\dots<q_k$. In this notation,

$$r_k=\sum_{\overline{p}} t_{p_1}\cdots t_{p_k}\sum_{\overline{q}} D^2_{\overline{p}\overline{q}}. \tag{44}$$

Since $d_{kl}=0$ for $k>l$ and $d_{kk}=1$, we have in particular $D_{\overline{p}\overline{p}}=1$, and in the case $p_l=n-k+l$ $(l=1,\dots,k)$, $D_{\overline{p}\overline{q}}=0$ for $\overline{q}\ne\overline{p}$. Let furthermore c_1, c_2, c_3, c_4 denote positive numbers only depending on n. By (44) we have the following estimate

$$t_{n-k+1}t_{n-k+2}\cdots t_n \;\le\; r_k \;<\; c_1 t_{n-k+1}t_{n-k+2}\cdots t_n.$$

On the other hand, trivially

$$\tau_{n-k+1}\tau_{n-k+2}\cdots\tau_n \;\le\; r_k \;\le\; \binom{n}{k}\tau_{n-k+1}\tau_{n-k+2}\cdots\tau_n. \tag{45}$$

Therefore, τ_k and t_k are of the same order for each $k=1,\dots,n$, so

$$\tau_k<c_2 t_k,\quad t_k<c_2\tau_k\quad (k=1,\dots,n). \tag{46}$$

Let $\rho_1,\dots,\rho_n$ be the eigenvalues of $\boldsymbol{R}[\boldsymbol{U}]$, in ascending order, and let $s_1,\dots,s_n$ be their elementary symmetric polynomials. If the subdeterminants of $\boldsymbol{U}$ are denoted by $U_{\overline{p}\overline{q}}$ in accordance with the convention above, then from $\boldsymbol{R}[\boldsymbol{U}]=\boldsymbol{T}[\boldsymbol{D}\boldsymbol{U}]$ we obtain the formula

$$s_k=\sum_{\overline{p}} t_{p_1}\cdots t_{p_k}\sum_{\overline{q}}\Big(\sum_{\overline{v}} D_{\overline{p}\overline{v}}U_{\overline{v}\overline{q}}\Big)^2$$

and from this

$$s_k\ge t_{n-k+1}t_{n-k+2}\cdots t_n\sum_{\overline{q}} U^2_{\overline{q}}, \tag{47}$$

where the $U_{\overline{q}}=U_{\overline{p}\overline{q}}$ are formed with the last k rows of $\boldsymbol{U}$, that is, with $p_l=n-k+l$ $(l=1,\dots,k)$. Since these rows are linearly independent, at least one of the subdeterminants is $U_{\overline{q}}\ne 0$, so that also

$$\sum_{\overline{q}} U_{\overline{q}}^2 = u_k \geq 1 \quad (k = 1, \ldots, n). \tag{48}$$

On the other hand, we again have

$$s_k \leq \binom{n}{k} \rho_{n-k+1} \rho_{n-k+2} \cdots \rho_n. \tag{49}$$

For short, we write

$$\log(\rho_k) = \lambda_k, \quad \log(\tau_k) = \mu_k,$$
$$\lambda_k + \lambda_{k+1} + \cdots + \lambda_n = \alpha_k, \quad \mu_k + \mu_{k+1} + \cdots + \mu_n = \beta_k$$

for $k = 1, \ldots, n$. As $|\boldsymbol{R}| = |\boldsymbol{R}[\boldsymbol{U}]|$, also $\alpha_1 = \beta_1$, and we define $\alpha_{n+1} = \beta_{n+1} = 0$. With $d(\boldsymbol{R}) = \mu$, $d(\boldsymbol{R}[\boldsymbol{U}]) = \lambda$ we obtain

$$\mu^2 = \sum_{k=1}^{n} \mu_k^2, \quad \lambda^2 = \sum_{k=1}^{n} \lambda_k^2,$$
$$\lambda^2 - \mu^2 = \sum_{k=1}^{n} (\lambda_k - \mu_k)(\lambda_k + \mu_k) = \sum_{k=1}^{n} \big((\alpha_k - \beta_k) - (\alpha_{k+1} - \beta_{k+1})\big)(\lambda_k + \mu_k),$$

and

$$\lambda - \mu = \sum_{k=2}^{n} (\alpha_k - \beta_k) \frac{(\lambda_k - \lambda_{k-1}) + (\mu_k - \mu_{k-1})}{\lambda + \mu}. \tag{50}$$

This still holds in the trivial case $\lambda = \mu = 0$, if every fraction is replaced by 1. Moreover, by (46), (47) and (49) we have

$$u_k \tau_{n-k+1} \tau_{n-k+2} \cdots \tau_n < c_3 \rho_{n-k+1} \rho_{n-k+2} \cdots \rho_n,$$

and hence

$$\alpha_{n-k+1} - \beta_{n-k+1} > \log(u_k) - c_4 \geq -c_4 \quad (k = 1, \ldots, n). \tag{51}$$

By

$$\lambda_k \geq \lambda_{k-1}, \quad \mu_k \geq \mu_{k-1} \quad (k = 2, \ldots, n), \tag{52}$$

the fractions in (50) are all non-negative, so (51) yields

$$\lambda - \mu > -c_4 \frac{\lambda_n - \lambda_1 + \mu_n - \mu_1}{\lambda + \mu} \geq -c_4 \sqrt{2}, \tag{53}$$

which proves the claim of Theorem 4.

Furthermore it needs to be investigated if the bound appearing in the theorem can be chosen independently of $\boldsymbol{P}$ if $\boldsymbol{P}$ is also contained in R. However, this seems to be a more difficult question.

11 Asymptotes

In the notation from Section 8 let $\boldsymbol{H} = \boldsymbol{T}^{\rho}[\boldsymbol{C}]$ with $\boldsymbol{T} \neq \boldsymbol{E}$ be a given geodesic line in P. Here, $\boldsymbol{T}$ is a diagonal matrix with positive diagonal elements $t_1, \ldots, t_n$, ordered increasingly. The multiplicities of the t_k shall correspond to the type $\{\kappa\}$, that is

$$t_k = t_l \quad (\kappa_{p-1} < k \leq \kappa_p;\ \kappa_{p-1} < l \leq \kappa_p;\ p = 1, \ldots, r),$$
$$t_k < t_l \quad (k \leq \kappa_p < l;\ p = 1, \ldots, r-1).$$

We then say that the geodesic line is of *type* $\{\kappa\}$. As $\boldsymbol{T} \neq \boldsymbol{E}$, the empty type is excluded.

For a point $\boldsymbol{P}$ in P, let the distance $d(\boldsymbol{P}, \boldsymbol{H})$ and the reduced distance $f(\boldsymbol{P}, \boldsymbol{H})$ be given. The geodesic line will be called an *asymptote* if the difference $d(\boldsymbol{P}, \boldsymbol{H}) - f(\boldsymbol{P}, \boldsymbol{H})$ remains bounded for $\rho \to \infty$. From the definition of the reduced distance it is clear that this distance is always non-negative. It furthermore follows from the triangle inequality that the definition of an asymptote is independent of the choice of $\boldsymbol{P}$. So we might as well choose $\boldsymbol{P} = \boldsymbol{E}$. Similarly, with $\boldsymbol{H} = \boldsymbol{T}^{\rho}[\boldsymbol{C}]$, also $\boldsymbol{H} = \boldsymbol{T}^{\rho}[\boldsymbol{CU}]$ for arbitrary unimodular $\boldsymbol{U}$ is an asymptote. So unimodular transformations preserve the asymptote property.

For any $\rho > 0$ let a unimodular $\boldsymbol{V} = \boldsymbol{V}(\rho)$ be chosen such that $\boldsymbol{H}[\boldsymbol{V}]$ lies in R. Then $f(\boldsymbol{P}, \boldsymbol{H}) = f(\boldsymbol{P}, \boldsymbol{H}[\boldsymbol{V}])$, and by Theorem 4, the difference $d(\boldsymbol{P}, \boldsymbol{H}[\boldsymbol{V}]) - f(\boldsymbol{P}, \boldsymbol{H}[\boldsymbol{V}])$ remains bounded for $\rho \to \infty$. Further, considering the triangle inequality, the asymptote condition can be replaced by the simpler requirement that $d(\boldsymbol{H}) - d(\boldsymbol{H}[\boldsymbol{V}])$ is bounded from above. Here, the notion of reduced distance no longer appears.

Theorem 5. *The geodesic line $\boldsymbol{H} = \boldsymbol{T}^{\rho}[\boldsymbol{C}]$ of type $\{\kappa\}$ is an asymptote if and only if there exists a unimodular matrix $\boldsymbol{W}$, such that matrix $\boldsymbol{CW}$ splits of type $\{\kappa\}$.*

It is easy to see that the condition mentioned in the theorem is sufficient, where we only need to consider the case $\boldsymbol{W} = \boldsymbol{E}$ due to the aforementioned invariance property. But then $\boldsymbol{H} = \boldsymbol{T}^{\rho}[\boldsymbol{C}]$ converges to a boundary point α of type $\{\kappa\}$ for $\rho \to \infty$. On the other hand, $\overline{R}$ is compact and $\boldsymbol{H}[\boldsymbol{V}]$ lies in R, so that by the result from Section 6 the unimodular matrix $\boldsymbol{V}(\rho)$ remains bounded for $\rho \to \infty$. Since

$$d(\boldsymbol{E}, \boldsymbol{H}) - d(\boldsymbol{E}, \boldsymbol{H}[\boldsymbol{V}]) = d(\boldsymbol{E}[\boldsymbol{V}], \boldsymbol{H}[\boldsymbol{V}]) - d(\boldsymbol{E}, \boldsymbol{H}[\boldsymbol{V}]) \leq d(\boldsymbol{E}, \boldsymbol{E}[\boldsymbol{V}]),$$

the difference $d(\boldsymbol{H}) - d(\boldsymbol{H}[\boldsymbol{V}])$ is thus indeed bounded from above.

The proof of necessity of the condition lies somewhat deeper. If $\rho_1, \ldots, \rho_n$ are the eigenvalues of $\boldsymbol{H} = \boldsymbol{T}^{\rho}[\boldsymbol{C}]$ in increasing order, and $s_1, \ldots, s_n$ are their elementary symmetric polynomials, then in analogy to (44) we have the relation

$$s_k = \sum_{\overline{p}} (t_{p_1} \cdots t_{p_k})^{\rho} \sum_{\overline{q}} C_{\overline{pq}}^2, \tag{54}$$

where $C_{\overline{pq}}$ runs through the k-row subdeterminants of $\boldsymbol{C}$. Since any k rows of $\boldsymbol{C}$ are linearly independent, any of the inner sums of (54) is positive. This implies the asymptotic relation

$$s_k \sim \gamma_k (t_{n-k+1} t_{n-k+2} \cdots t_n)^{\rho} \quad (k = 1, \ldots, n;\ \rho \to \infty)$$

for certain positive quantities γ_k that are independent of ρ. By using (45) we find

$$\lambda_k = \log(\rho_k) = \rho \log(t_k) + \mathrm{O}(1) \quad (k = 1, \ldots, n), \tag{55}$$

$$\lambda_k - \lambda_{k-1} = \rho \log\left(\frac{t_k}{t_{k-1}}\right) + \mathrm{O}(1) \quad (k = 2, \ldots, n). \tag{56}$$

Let $\tau_1, \ldots, \tau_n$ be the eigenvalues of $\boldsymbol{H}[\boldsymbol{V}] = \boldsymbol{T}^\rho[\boldsymbol{CV}]$ in increasing order, and let $r_1, \ldots, r_n$ be their elementary symmetric polynomials. Furthermore, let again

$$\log(\tau_k) = \mu_k, \quad \sum_{l=k}^{n} \lambda_l = \alpha_k, \quad \sum_{l=k}^{n} \mu_l = \beta_k,$$

$$\mu^2 = \sum_{k=1}^{n} \mu_k^2, \quad \lambda^2 = \sum_{k=1}^{n} \lambda_k^2,$$

so that $\lambda = d(\boldsymbol{H})$ and $\mu = d(\boldsymbol{H}[\boldsymbol{V}])$. The remaining positive quantities $c_5, \ldots, c_{18}$ depend on n as well as on $\boldsymbol{T}$ and $\boldsymbol{C}$, but not on ρ. From

$$d(\boldsymbol{H}, \boldsymbol{C}'\boldsymbol{C})^2 = \rho^2 \sum_{k=1}^{n} \log(t_k)^2$$

we obtain

$$\lambda < c_5 \rho + c_6. \tag{57}$$

We make a connection to the formulas from the previous section by setting $\boldsymbol{H}[\boldsymbol{V}] = \boldsymbol{R}$, $\boldsymbol{V}^{-1} = \boldsymbol{U}$, $\boldsymbol{H} = \boldsymbol{R}[\boldsymbol{U}]$. According to (51) and (53) we then have

$$\alpha_k - \beta_k > -c_4 \quad (k = 1, \ldots, n) \tag{58}$$

and

$$\lambda - \mu > -c_4 \sqrt{2}. \tag{59}$$

Now, if $\boldsymbol{H} = \boldsymbol{T}^\rho[\boldsymbol{C}]$ is an asymptote, then also

$$\lambda - \mu < c_7 \quad (\rho \to \infty). \tag{60}$$

From (50), (52), (58) and (60) follows

$$c_7 > (\alpha_k - \beta_k) \frac{\lambda_k - \lambda_{k-1}}{\lambda + \mu} - c_4 \sqrt{2} \quad (k = 2, \ldots, n),$$

and from this it further follows with (56), (57) and (59) the estimate

$$\alpha_k - \beta_k < c_8 \quad (k - 1 = \kappa_1, \ldots, \kappa_{r-1}), \tag{61}$$

since precisely for these values of $k-1$ we have the ratio $t_k / t_{k-1} > 1$. According to (51) it also holds for these values of $k-1$ that

$$\log(u_{n-k+1}) < c_9. \tag{62}$$

By the meaning of u_k given by (48), (62) states the boundedness of all subdeterminants formed from the last $n - \kappa_p$ rows of the matrix $\boldsymbol{U} = \boldsymbol{V}^{-1}$, namely for $p = 1, \ldots, r-1$.

First choose $p = r-1$. With our earlier notation, $n-\kappa_r = j_r$. Let $\boldsymbol{M}$ be the matrix formed by the last j_r rows of $\boldsymbol{U}$. Now we wish to determine a bounded unimodular matrix $\boldsymbol{W}$ such that

$$\boldsymbol{M}\boldsymbol{W} = (0\ \boldsymbol{U}_r) \tag{63}$$

for a matrix $\boldsymbol{U}_r$ with j_r rows. In order to do this conveniently by induction, we will more generally only assume that $\boldsymbol{M}$ is integral and of rank $j_r = k$. Moreover, we assume that all k-row subdeterminants of $\boldsymbol{M}$ are bounded. For $n = k$, $\boldsymbol{W} = \boldsymbol{E}$ is the desired matrix. Use induction on n for fixed k. Without loss of generality we may assume here that the matrix $\boldsymbol{M}_1$ obtained from $\boldsymbol{M}$ by omitting the last column $\boldsymbol{q}$ also has rank k, as this can be guaranteed in the case $k < n$ by multiplication with a suitable permutation matrix $\boldsymbol{W}$. The k-row subdeterminants of $\boldsymbol{M}_1$ are bounded, hence by induction we can find a bounded unimodular matrix $\boldsymbol{W}_1$ such that

$$\boldsymbol{M}_1\boldsymbol{W}_1 = (0\ \boldsymbol{M}_2),$$

where $\boldsymbol{M}_2$ is quadratic with k rows. Then also

$$\boldsymbol{M}\begin{pmatrix}\boldsymbol{M}_1 & 0\\ 0 & 1\end{pmatrix} = (0\ \boldsymbol{M}_2\ \boldsymbol{q}).$$

The matrix $(\boldsymbol{M}_2\ \boldsymbol{q}) = \boldsymbol{M}_3$ is integral. It has k rows, $k+1$ columns and rank k, its subdeterminants of degree k are all bounded by the Laplace theorem, as this holds for $\boldsymbol{M}$ itself. If we form the column $\boldsymbol{p}$ from these $k+1$ subdeterminants of $\boldsymbol{M}_3$, then $\boldsymbol{p} \neq 0$, integral and bounded. If t denotes the greatest common divisor of the elements of $\boldsymbol{p}$, then column $t^{-1}\boldsymbol{p} = \boldsymbol{r}$ is primitive and can thus be augmented to a bounded unimodular matrix $\boldsymbol{W}_2$. As $\boldsymbol{M}_3\boldsymbol{r} = 0$, then accordingly

$$\boldsymbol{M}_3\boldsymbol{W}_2 = (0\ *)$$

and the desired matrix is

$$\boldsymbol{W} = \begin{pmatrix}\boldsymbol{W}_1 & 0\\ 0 & 1\end{pmatrix}\begin{pmatrix}\boldsymbol{E} & 0\\ 0 & \boldsymbol{W}_2\end{pmatrix}.$$

According to (63),

$$\boldsymbol{U}\boldsymbol{W} = \begin{pmatrix}\boldsymbol{U}_0 & *\\ 0 & \boldsymbol{U}_r\end{pmatrix} \tag{64}$$

with unimodular $\boldsymbol{U}_0$ and $\boldsymbol{U}_r$. Now choose $p = r-2$, that is, $n-\kappa_p = n-\kappa_{r-2} = j_{r-1}+j_r = k$. As all of the k-row subdeterminants formed from the last k rows of $\boldsymbol{U}$ are bounded, the Laplace expansion theorem implies the same for the matrix $\boldsymbol{U}\boldsymbol{W}$ instead of $\boldsymbol{U}$. Because of the decomposition (64) this means that all subdeterminants of degree j_{r-1} formed from the j_{r-1} last rows of $\boldsymbol{U}_0$ are bounded as well. From this it follows by induction on decreasing values of r that there exists a bounded unimodular matrix $\boldsymbol{W}$ such that the product $\boldsymbol{U}\boldsymbol{W} = \boldsymbol{L}$ belongs to Γ_κ. Note that since $\boldsymbol{U}^{-1} = \boldsymbol{V} = \boldsymbol{V}(\rho)$, the matrix $\boldsymbol{W}$ still depends on ρ.

We will now prove that the thus constructed matrix $\boldsymbol{W}$ has the properties claimed in Theorem 5. Set $\boldsymbol{C}\boldsymbol{W} = \boldsymbol{B}$ and compute the principal subdeterminants of the matrix

$$\boldsymbol{R}[\boldsymbol{L}] = \boldsymbol{H}[\boldsymbol{W}] = \boldsymbol{T}^{\rho}[\boldsymbol{C}\boldsymbol{W}] = \boldsymbol{T}^{\rho}[\boldsymbol{B}] \tag{65}$$

of degrees $k = \kappa_1, \ldots, \kappa_{r-1}$. These principal subdeterminants equal the corresponding ones for $\boldsymbol{R}$, as $\boldsymbol{L}$ belongs to Γ_κ. According to (46) with the corresponding meaning of the t_k, the principal subdeterminants are precisely of the order of the products $\tau_1 \cdots \tau_k$. By (65),

$$c_{10}\tau_1 \cdots \tau_k > \sum_{\overline{p}} (t_{p_1} \cdots t_{p_k})^{\rho} B_{\overline{p}}^2 \quad (k = \kappa_1, \ldots, \kappa_{r-1}), \tag{66}$$

where $B_{\overline{p}}$ denotes the subdeterminant of $\boldsymbol{B}$ that is formed from the first k columns and the rows with indices $p_1, \ldots, p_k$. Now

$$\sum_{l=1}^{k} (\log(\tau_l) - \log(\rho_l)) = \sum_{l=1}^{k} (\mu_l - \lambda_l) = (\beta_1 - \beta_{k+1}) - (\alpha_1 - \alpha_{k+1}) = \alpha_{k+1} - \beta_{k+1},$$

hence by (61) also

$$\log\left(\frac{\tau_1 \cdots \tau_k}{\rho_1 \cdots \rho_k}\right) < c_{11} \quad (k = \kappa_1, \ldots, \kappa_{r-1};\ \rho \to \infty).$$

With (55) this implies

$$\tau_1 \cdots \tau_k < c_{12}(t_1 \cdots t_k)^{\rho}$$

for the same values of k, and comparison with (66) finally yields

$$B_{\overline{p}}^2 < c_{13}\left(\frac{t_1 \cdots t_k}{t_{p_1} \cdots t_{p_k}}\right)^{\rho}.$$

If we take into account that $t_{k+1} > t_k$ holds and that there are only finitely many possibilities for $\boldsymbol{B}$, then with $\rho \to \infty$ the vanishing of all $B_{\overline{p}}$ follows, with the exception of the case $p_1 = 1$, $p_2 = 2$, …, $p_k = k$. Since $|\boldsymbol{B}| \neq 0$, the first k columns of $\boldsymbol{B}$ are linearly independent. The vanishing of the $B_{\overline{p}}$ then implies that all elements in $\boldsymbol{B}$ that appear in the first k rows below the k-th column are 0, for $k = \kappa_1, \ldots, \kappa_{r-1}$. Hence $\boldsymbol{B} = (\boldsymbol{B}_{pq})$ with boxes $\boldsymbol{B}_{pq}$ of j_p rows and j_q columns $(p, q = 1, \ldots, r)$ and $\boldsymbol{B}_{pq} = 0$ for $p > q$. This concludes the proof of the theorem.

We will now draw some conclusions from the previously obtained results. For this it is convenient to replace the matrix $\boldsymbol{W}$ appearing in Theorem 5 by $\boldsymbol{W}^{-1}$, so that the condition for $\boldsymbol{C}$ mentioned there becomes $\boldsymbol{C} = \boldsymbol{B}\boldsymbol{W}$, where $\boldsymbol{B}$ splits of type $\{\kappa\}$. The matrix $\boldsymbol{W}$ is clearly only determined up to a factor on the left from Γ_κ, and for any choice of such a factor, $\boldsymbol{H}[\boldsymbol{W}^{-1}] = \boldsymbol{T}^{\rho}[\boldsymbol{B}]$ converges to a boundary point of type $\{\kappa\}$ for $\rho \to \infty$.

Conversely, assume now that for any geodesic line $\boldsymbol{H} = \boldsymbol{T}^{\rho}[\boldsymbol{C}]$ of type $\{\kappa\}$ and a suitable unimodular matrix $\boldsymbol{U}$ the equivalent geodesic line $\boldsymbol{H}[\boldsymbol{U}^{-1}] = \boldsymbol{T}^{\rho}[\boldsymbol{C}\boldsymbol{U}^{-1}]$ converges to a boundary point β of type $\{\lambda\}$ for $\rho \to \infty$. If we again choose $\boldsymbol{V} = \boldsymbol{V}(\rho)$ such that $\boldsymbol{H}[\boldsymbol{U}^{-1}\boldsymbol{V}]$ is reduced, then by the result of Section 6, the matrix $\boldsymbol{V}$ is bounded and lies in Γ_λ for sufficiently large ρ. But then the difference

$$d(\boldsymbol{H}[\boldsymbol{U}^{-1}]) - d(\boldsymbol{H}[\boldsymbol{U}^{-1}\boldsymbol{V}]) \le d(\boldsymbol{E}[\boldsymbol{V}])$$

is also bounded, and therefore the geodesic line $\boldsymbol{H}[\boldsymbol{U}^{-1}] = \boldsymbol{T}^{\rho}[\boldsymbol{C}\boldsymbol{U}^{-1}]$ is an asymptote. So by Theorem 5 there exists an unimodular matrix $\boldsymbol{W}_1$ such that $\boldsymbol{H}[\boldsymbol{U}^{-1}\boldsymbol{W}_1^{-1}] = \boldsymbol{T}^{\rho}[\boldsymbol{C}\boldsymbol{U}^{-1}\boldsymbol{W}_1^{-1}]$ converges to a boundary point α of type $\{\kappa\}$ for $\rho \to \infty$. By Theorem 1 we then have $\{\lambda\} = \{\kappa\}$, $\boldsymbol{W}_1 \in \Gamma_\kappa$ and $\beta = \alpha_{\boldsymbol{W}_1}$. If $\boldsymbol{H} = \boldsymbol{T}^{\rho}[\boldsymbol{C}]$ is again the asymptote from Theorem 5, then in particular we can choose $\boldsymbol{W} = \boldsymbol{W}_1\boldsymbol{U}$. Thus $\boldsymbol{W}$ and $\boldsymbol{U}$ belong to the same right coset of Γ_κ in Γ.

In particular, assume now that the geodesic line $\boldsymbol{H}(\rho) = \boldsymbol{T}^{\rho}[\boldsymbol{C}]$ converges to a boundary point α of type $\{\kappa\}$ for $\rho \to \infty$. Then $\boldsymbol{C}$ can be decomposed into boxes with $\boldsymbol{C}_{pq} = 0$ for $p > q$. We set $\boldsymbol{C}_{pp}^{-1}\boldsymbol{C}_{pq} = \boldsymbol{D}_{pq}$ and $\boldsymbol{C}'_{pp}\boldsymbol{C}_{pp} = b_p\boldsymbol{R}_p$ with scalar $b_p > 0$ and $|\boldsymbol{R}_p| = 1$. Furthermore let $t_k = \theta_p$ $(\kappa_{p-1} < l \le \kappa_p;\ p = 1,\dots,r)$ and choose a normalization

$$\sum_{k=1}^{n} \log(t_k)^2 = 1 \tag{67}$$

such that $\rho = s$ is the arc length of the geodesic line $\boldsymbol{H} = \boldsymbol{H}(\rho)$. In addition set $b_p\theta_p^s = \tau_p$ $(p = 1,\dots,r)$ and $\tau_p/\tau_{p+1} = v_p$ $(p = 1,\dots,r-1)$. Then the $v_p, \boldsymbol{R}_p, \boldsymbol{D}_{pq}$ $(p < q)$ are just the normal coordinates of type $\{\kappa\}$ for $\boldsymbol{H}$, and the geodesic line runs through the geodesic plane $G(\alpha)$ that was introduced in Section 8. For $s = 0$ we have $\tau_p = b_p$. Hence, for given $\theta_1,\dots,\theta_r$, in ascending order, there exists through any point of P exactly one geodesic line of type $\{\kappa\}$ that converges to a boundary point. By varying the θ_p we obtain a family with $r-2$ parameters, since in addition to condition (67) also the condition $t_1 \cdots t_n = 1$ must be taken into account. It thus is clear how, for a given boundary point α, all geodesic lines converging to α can be determined. If we also prescribe the values $\theta_1,\dots,\theta_r$, then they depend on $r-2$ parameters.

12 Ideal points

It is useful to generalize the notion of a boundary point. Whereas for every unimodular $\boldsymbol{U}$ it makes sense to consider the map $\boldsymbol{H} \mapsto \boldsymbol{H}[\boldsymbol{U}]$ on the space P, this is no longer the case for $\overline{P}$. In fact, for a boundary point α of type $\{\kappa\}$ the corresponding image point $\alpha_{\boldsymbol{U}}$ has only been defined for $\boldsymbol{U} \in \Gamma_\kappa$.

Let $\boldsymbol{U}$ and $\boldsymbol{V}$ be two given unimodular matrices and let $\boldsymbol{H}$ run through a sequence in P such that $\boldsymbol{H}[\boldsymbol{U}^{-1}]$ and $\boldsymbol{H}[\boldsymbol{V}^{-1}]$ converge to boundary points α and β. According to Theorem 1, α and β are of the same type $\{\kappa\}$, and we have $\boldsymbol{U}\boldsymbol{V}^{-1} = \boldsymbol{W} \in \Gamma_\kappa$ as well as $\alpha_{\boldsymbol{W}} = \beta$. Thus $\boldsymbol{V}$ and $\boldsymbol{U} = \boldsymbol{W}\boldsymbol{V}$ lie in the same right coset C of Γ_κ in Γ. Conversely, $\boldsymbol{H}[\boldsymbol{U}^{-1}] \Rightarrow \alpha$ implies $\boldsymbol{H}[\boldsymbol{U}^{-1}\boldsymbol{W}] = \boldsymbol{H}[\boldsymbol{U}^{-1}][\boldsymbol{W}] \Rightarrow \alpha_{\boldsymbol{W}}$.

Now let α be a boundary point of type $\{\kappa\}$ and $\boldsymbol{U}$ in Γ. Let C denote the right coset of Γ_κ in Γ that contains $\boldsymbol{U}$. If C is the principal class, that is, $\boldsymbol{U} \in \Gamma_\kappa$, then $\alpha_{\boldsymbol{U}}$ is a well-defined boundary point of type $\{\kappa\}$. On the other hand, if $\boldsymbol{U}$ is not an element of Γ_κ, then we introduce $\alpha_{\boldsymbol{U}}$ as a new symbol and call this an *ideal point* of the class C of type $\{\kappa\}$. In this sense, the boundary points are the ideal points of the principal class. A sequence $\boldsymbol{H} \in P$ will be called *convergent to the ideal point* $\alpha_{\boldsymbol{U}}$ precisely if $\boldsymbol{H}[\boldsymbol{U}^{-1}] \Rightarrow \alpha$ holds in the usual sense. We write $\boldsymbol{H} \Rightarrow \alpha_{\boldsymbol{U}}$. If also $\boldsymbol{H}[\boldsymbol{V}^{-1}] \Rightarrow \beta$ for a fixed unimodular $\boldsymbol{V}$, then $\boldsymbol{H} \Rightarrow \beta_{\boldsymbol{V}}$ and we set $\alpha_{\boldsymbol{U}} = \beta_{\boldsymbol{V}}$. On the other hand we have $\boldsymbol{U} = \boldsymbol{W}\boldsymbol{V}$ with $\boldsymbol{W} \in \Gamma_\kappa$ and $\alpha_{\boldsymbol{W}} = \beta$, hence $\beta_{\boldsymbol{V}} = (\alpha_{\boldsymbol{W}})_{\boldsymbol{V}} = \alpha_{\boldsymbol{W}\boldsymbol{V}}$, and $\boldsymbol{V}$ also lies in the class C. Conversely, it follows from $\boldsymbol{U} = \boldsymbol{W}\boldsymbol{V}$ with $\boldsymbol{W} \in \Gamma_\kappa$ and $\boldsymbol{H}[\boldsymbol{U}^{-1}] \Rightarrow \alpha$ that $\boldsymbol{H}[\boldsymbol{V}^{-1}] \Rightarrow \alpha_{\boldsymbol{W}} = \beta$, that is, $\alpha_{\boldsymbol{U}} = \beta_{\boldsymbol{V}}$. This definition of equality identifies only certain ideal points in the same class and is moreover independent of the choice of the sequence $\boldsymbol{H}$, for $\boldsymbol{P}[\boldsymbol{U}^{-1}] \Rightarrow \alpha$ also implies $\boldsymbol{P}[\boldsymbol{V}^{-1}] = \boldsymbol{P}[\boldsymbol{U}^{-1}][\boldsymbol{W}] \Rightarrow \alpha_{\boldsymbol{W}} = \beta$.

If $\alpha_{\boldsymbol{U}}$ and $\gamma_{\boldsymbol{V}}$ are any two ideal points in the same class, then $\boldsymbol{U} = \boldsymbol{W}\boldsymbol{V}$ with $\boldsymbol{W} \in \Gamma_\kappa$, hence $\alpha_{\boldsymbol{W}} = \beta$ is a boundary point of type $\{\kappa\}$ and $\alpha_{\boldsymbol{U}} = \beta_{\boldsymbol{V}}$. In this manner we obtain the ideal points of class C uniquely in the form $\alpha_{\boldsymbol{U}}$ with a fixed class representative $\boldsymbol{U}$, where α runs through the set P_κ of boundary points of type $\{\kappa\}$. If we fix the class representative, then we can introduce for $\alpha_{\boldsymbol{U}}$ the coordinates of α. Moreover, it is clear how to transfer the notion of a neighborhood; this is of course independent of the choice of $\boldsymbol{U}$ in C. Due to the invariance property of distances by unimodular transformations, $\alpha_{\boldsymbol{U}}$ and $\gamma_{\boldsymbol{V}}$ have a certain finite distance for which

$$d(\alpha_{\boldsymbol{U}}, \gamma_{\boldsymbol{V}}) = d(\beta, \gamma)$$

holds. On the other hand, if $\alpha_{\boldsymbol{U}}$ and $\gamma_{\boldsymbol{V}}$ are not in the same class and possibly α and γ even of different type, then $\boldsymbol{H} \Rightarrow \alpha_{\boldsymbol{U}}$, $\boldsymbol{P} \Rightarrow \gamma_{\boldsymbol{V}}$ and Theorem 2 imply that in the limit the distance

$$d(\boldsymbol{P}, \boldsymbol{H}) = d(\boldsymbol{P}[\boldsymbol{V}^{-1}], \boldsymbol{H}[\boldsymbol{U}^{-1}][\boldsymbol{U}\boldsymbol{V}^{-1}])$$

grows beyond all bounds. The distance of $\alpha_{\boldsymbol{U}}$ and $\gamma_{\boldsymbol{V}}$ is then infinite.

If $\boldsymbol{H} \Rightarrow \alpha_{\boldsymbol{U}}$, then $\boldsymbol{H}[\boldsymbol{V}] \Rightarrow \alpha_{\boldsymbol{U}\boldsymbol{V}}$ for arbitrary unimodular $\boldsymbol{V}$. This makes evident the definition $(\alpha_{\boldsymbol{U}})_{\boldsymbol{V}} = \alpha_{\boldsymbol{U}\boldsymbol{V}}$, where we observe that this is compatible with the above identification. In this way, the map $\boldsymbol{H} \mapsto \boldsymbol{H}[\boldsymbol{V}]$ has been extended to all ideal points, namely the classes of ideal points are permuted via the assignment $C \mapsto C\boldsymbol{V}$, whereas the types are preserved. Let P^* denote the space obtained from $\overline{P}$ by adding all ideal points. Then Γ acts as a transformation group on all of P^* and is discontinuous throughout. Moreover, $\overline{R}$ is a fundamental domain for Γ in P^*.

If $\boldsymbol{H} = \boldsymbol{T}^\rho[\boldsymbol{C}]$ is an asymptote of type $\{\kappa\}$, then there exists a uniquely determined right coset C of Γ_κ in Γ, such that for a representatives $\boldsymbol{U}$ of C and $\rho \to \infty$ the equivalent geodesic line $\boldsymbol{H}[\boldsymbol{U}^{-1}] = \boldsymbol{T}^\rho[\boldsymbol{C}\boldsymbol{U}^{-1}]$ tend to a boundary point α of type $\{\kappa\}$. Hence $\boldsymbol{H} \Rightarrow \alpha_{\boldsymbol{U}}$. We transfer the distinct classes to the asymptotes. Then every asymptote passes through a certain ideal point of the same class. Moreover, it follows from the remarks at the end of the previous section that for any given class C with fixed normalized eigenvalues $\theta_1, \ldots, \theta_r$ there exists through every point of P exactly one asymptote in this class, whereas we obtain a family with $r-2$ parameters if we allow the eigenvalues to vary. The asymptotes passing through a given prescribed ideal point depend on the same number of parameters if the eigenvalues are given.

13 The homogenous space of the orthogonal group

As is well-known, we owe the idea on how to include the reduction theory of indefinite quadratic forms into that of definite ones to Hermite. Namely, if $\boldsymbol{S}[\boldsymbol{x}]$ is a non-degenerate indefinite quadratic form with real coefficients in n variables, then consider all real linear substitutions $\boldsymbol{y} = \boldsymbol{C}\boldsymbol{x}$ that transform $\boldsymbol{S}[\boldsymbol{x}]$ into a sum of positive and negative squares of the variables $y_1, \ldots, y_n$, and with this obtain the positive definite quadratic form $\boldsymbol{y}'\boldsymbol{y} = \boldsymbol{H}[\boldsymbol{x}]$ whose matrix is $\boldsymbol{H} = \boldsymbol{C}'\boldsymbol{C}$. If there is a reduced matrix among these positive matrices $\boldsymbol{H}$, then $\boldsymbol{S}$ is also called *reduced.*

Let

$$\boldsymbol{S}[\boldsymbol{x}] = (y_1^2 + \cdots + y_m^2) - (y_{m+1}^2 + \cdots + y_n^2) = \boldsymbol{S}_0[\boldsymbol{y}],$$

where $\boldsymbol{S}_0$ is the diagonal matrix with m diagonal elements 1 and $n-m$ diagonal elements -1, and m, $n-m$ the signature of $\boldsymbol{S}$. Then $\boldsymbol{S}[\boldsymbol{C}^{-1}] = \boldsymbol{S}_0$ and $\boldsymbol{S}_0^2 = \boldsymbol{E}$, from which we deduce the relation

$$\boldsymbol{S}\boldsymbol{H}^{-1}\boldsymbol{S}\boldsymbol{H}^{-1} = \boldsymbol{E}, \quad \boldsymbol{H} > 0 \tag{68}$$

by elimination of $\boldsymbol{C}$. Conversely, it is easy to see by a principal axis transformation that for every solution $\boldsymbol{H}$ of (68) there exists a real matrix $\boldsymbol{C}$ with $\boldsymbol{H} = \boldsymbol{C}'\boldsymbol{C}$, $\boldsymbol{S} = \boldsymbol{S}_0[\boldsymbol{C}]$. Let $H = H(\boldsymbol{S})$ be the space of all real solutions to (68) and $\Omega(\boldsymbol{S})$ the *orthogonal group* of $\boldsymbol{S}$ that consists of all real solutions $\boldsymbol{G}$ of $\boldsymbol{S}[\boldsymbol{G}] = \boldsymbol{S}$. Clearly, H is mapped to itself by the maps $\boldsymbol{H} \mapsto \boldsymbol{H}[\boldsymbol{G}]$. If in addition $\boldsymbol{S} = \boldsymbol{S}_0[\boldsymbol{C}_0]$ for a fixed real $\boldsymbol{C}_0$ and also $\boldsymbol{S} = \boldsymbol{S}_0[\boldsymbol{C}]$, then $\boldsymbol{C}_0^{-1}\boldsymbol{C} = \boldsymbol{G} \in \Omega(\boldsymbol{S})$, and vice versa. With $\boldsymbol{C}_0'\boldsymbol{C}_0 = \boldsymbol{H}_0$ it then follows that $\boldsymbol{C} = \boldsymbol{C}_0\boldsymbol{G}$ and $\boldsymbol{H} = \boldsymbol{C}'\boldsymbol{C} = \boldsymbol{H}_0[\boldsymbol{G}]$. Hence the maps $\boldsymbol{H} \mapsto \boldsymbol{H}[\boldsymbol{G}]$ on H are transitive. The subgroup $\Omega(\boldsymbol{S}, \boldsymbol{H}_0)$ of $\Omega(\boldsymbol{S})$ defined by $\boldsymbol{H}_0 = \boldsymbol{H}_0[\boldsymbol{G}]$ is compact, as $\boldsymbol{H}_0 > 0$. Moreover, $\Omega(\boldsymbol{S}, \boldsymbol{H}_0[\boldsymbol{G}]) = \boldsymbol{G}^{-1}\Omega(\boldsymbol{S}, \boldsymbol{H}_0)\boldsymbol{G}$ for $\boldsymbol{G} \in \Omega(\boldsymbol{S})$, and these conjugate subgroups yield all the maximal compact subgroups of $\Omega(\boldsymbol{S})$. Also, $\Omega(\boldsymbol{S}) = \boldsymbol{C}_0^{-1}\Omega(\boldsymbol{S}_0)\boldsymbol{C}_0$, and the map $\boldsymbol{H} \mapsto \boldsymbol{H}[\boldsymbol{C}_0]$ maps $H(\boldsymbol{S}_0)$ to $H(\boldsymbol{S})$. It may seem redundant to introduce $H(\boldsymbol{S})$ in addition to $H(\boldsymbol{S}_0)$, but it is convenient for studying the group of units of $\boldsymbol{S}$. Henceforth $H(\boldsymbol{S})$ is called the *homogeneous space* for the orthogonal group of $\boldsymbol{S}$. The points of the homogeneous space correspond to the right cosets of $\Omega(\boldsymbol{S}, \boldsymbol{H}_0)$ in $\Omega(\boldsymbol{S})$.

In analogy to Cayley's formula for rotations, the solutions of (68) have a parameter representation

$$\boldsymbol{H} = 2\boldsymbol{Z} - \boldsymbol{S}, \quad \boldsymbol{Z} = \boldsymbol{T}^{-1}[\boldsymbol{X}'\boldsymbol{S}], \quad \boldsymbol{T} = \boldsymbol{S}[\boldsymbol{X}] > 0 \tag{69}$$

for some real matrix $\boldsymbol{X}$ with n rows and m columns. By replacing $\boldsymbol{S}$ by $-\boldsymbol{S}$ if necessary, we may assume $m \le m - n$. In (69), $\boldsymbol{X}$ is determined by $\boldsymbol{H}$ only up to a real factor $\boldsymbol{R}$ on the right with m rows and $|\boldsymbol{R}| \ne 0$, and we may normalize $\boldsymbol{X}' = (\boldsymbol{E}\ *)$ with the m-row identity matrix, if the subdeterminant of the first m rows of $\boldsymbol{S}$ is not zero. The homogeneous space thus has $m(n-m)$ dimensions. For later purposes we need a parameter representation other than (69). Let $\kappa = \{\kappa_1, \ldots, \kappa_{r-1}\}$ be a type and let $\boldsymbol{S} = (\boldsymbol{S}_{pq})$ be the corresponding decomposition of $\boldsymbol{S}$ into boxes $\boldsymbol{S}_{pq}$ $(p, q = 1, \ldots, r)$ with j_p rows and j_q columns. We further assume that $\boldsymbol{S}_{pq} = 0$ for $p + q \le r$, so that

$$\boldsymbol{S}[\boldsymbol{x}] = \sum_{p+q>r} \boldsymbol{x}_p'\boldsymbol{S}_{pq}\boldsymbol{x}_q \tag{70}$$

and the box matrix $(\boldsymbol{S}_{pq})$ contains zero boxes above the anti-diagonal. For short, write $\boldsymbol{S}_p = \boldsymbol{S}_{pq}$ for $p + q = r + 1$. Then

$$|\boldsymbol{S}| = \pm|\boldsymbol{S}_1| \cdots |\boldsymbol{S}_r|, \quad |\boldsymbol{S}_p| \ne 0 \quad (p = 1, \ldots, r).$$

Note that $\boldsymbol{S}_{r-p+1} = \boldsymbol{S}_p'$. In particular, $\boldsymbol{S}_\nu$ is symmetric for odd $r = 2\nu - 1$.

For even r, certainly $m = n - m$ and thus $n = 2m$ is an even number. For odd $r = 2v - 1$, $\boldsymbol{S}_v$ has the same signature difference $n - 2m$ as $\boldsymbol{S}$. To avoid having to distinguish between odd and even r in what follows, we make the following changes to our notation. If $r = 2v - 2$ is even, what was κ_{r-p} $(p = 0,\ldots,v-1)$ so far will be denoted by κ_{r-p+1}, and then set $\kappa_{v-1} = \kappa_v$. After this change, κ_p is defined in both cases for $p = 0,\ldots,2v-1$ and $\kappa_p + \kappa_{2v-p-1} = n$. Such a type is called *symmetric*. Then $r = 2v - 1$ and $\kappa_{v-1} \le \kappa_v$, where equality may appear, and accordingly $j_v \ge 0$. If $j_v = 0$, then $\boldsymbol{S}_v$ is the empty matrix with 0 rows, and other matrices with j_v rows or columns are to be interpreted accordingly. Via the substitution

$$\boldsymbol{S}_p \boldsymbol{y}_{r-p+1} = \frac{1}{2}\boldsymbol{S}_{pp}\boldsymbol{x}_p + \sum_{q=r-p+1}^{p-1} \boldsymbol{S}_{pq}\boldsymbol{x}_q \quad (p = v+1,\ldots,r), \tag{71}$$

$$\boldsymbol{y}_p = \boldsymbol{x}_p \quad (p = v,\ldots,r)$$

we obtain

$$\boldsymbol{S}[\boldsymbol{x}] = \sum_{p=1}^{r} \boldsymbol{y}'_p \boldsymbol{S}_p \boldsymbol{y}_{r-p+1} = \overline{\boldsymbol{S}}[\boldsymbol{y}],$$

where the matrix $\overline{\boldsymbol{S}}$ is obtained from $\boldsymbol{S}$ by replacing all boxes $\boldsymbol{S}_{pq}$ with $p + q > r + 1$ by zeros. The matrix $\boldsymbol{B} = (\boldsymbol{B}_{pq})$ of the substitution $\boldsymbol{y} = \boldsymbol{B}\boldsymbol{x}$ given by (71) is a triangular matrix of type $\{\kappa\}$, and thus

$$\boldsymbol{S} = \overline{\boldsymbol{S}}[\boldsymbol{B}].$$

We now introduce normal coordinates of type $\{\kappa\}$ for $\boldsymbol{H}$ and set accordingly

$$\boldsymbol{H} = \overline{\boldsymbol{H}}[\boldsymbol{D}] \tag{72}$$

with

$$\overline{\boldsymbol{H}} = [\boldsymbol{H}_1,\ldots,\boldsymbol{H}_r], \quad \boldsymbol{D} = (\boldsymbol{D}_{pq}), \quad \boldsymbol{D}_{pp} = \boldsymbol{E}_p, \quad \boldsymbol{D}_{pq} = 0 \text{ for } p > q.$$

In addition, let

$$\boldsymbol{D}\boldsymbol{B}^{-1} = \boldsymbol{C}. \tag{73}$$

The condition for $\boldsymbol{H}$ in (68) can then be brought into the form

$$\overline{\boldsymbol{H}}[\boldsymbol{C}\overline{\boldsymbol{S}}^{-1}\boldsymbol{C}'\overline{\boldsymbol{S}}] = \overline{\boldsymbol{H}}^{-1}[\overline{\boldsymbol{S}}].$$

Here, $\boldsymbol{C}\overline{\boldsymbol{S}}^{-1}\boldsymbol{C}'\overline{\boldsymbol{S}}$ is again a triangular matrix of type $\{\kappa\}$, whereas $\overline{\boldsymbol{H}}^{-1}[\overline{\boldsymbol{S}}]$ is a diagonal matrix of this type. From the uniqueness of the generalized Jacobi decomposition it follows that

$$\overline{\boldsymbol{S}}[\boldsymbol{C}] = \overline{\boldsymbol{S}}, \quad \overline{\boldsymbol{H}} = \overline{\boldsymbol{H}}^{-1}[\overline{\boldsymbol{S}}], \tag{74}$$

with the additional condition $\boldsymbol{H} > 0$, that is, $\overline{\boldsymbol{H}} > 0$. This replaces equation (68) for $\boldsymbol{H}$ by the two equations (74), of which the first one contains only $\boldsymbol{C}$ and the second one contains only $\overline{\boldsymbol{H}}$. The second equation in (74) yields

$$\boldsymbol{H}_p = \boldsymbol{H}_p^{-1}[\boldsymbol{S}_p] \quad (p + q = r + 1). \tag{75}$$

Hence the $\boldsymbol{H}_p$ for $p = 1,\ldots,v-1$ can be chosen arbitrarily subject to the condition $\boldsymbol{H}_p > 0$, whereas for $j_v > 0$ the matrix $\boldsymbol{H}_v$ is to be taken from the homogeneous space

$H(\boldsymbol{S}_\nu)$. The $\boldsymbol{H}_p$ with $p = \nu+1,\dots,r$ are then given by (75). For $\boldsymbol{C}$, an arbitrary triangular matrix of type $\{\kappa\}$ in the group $\Omega(\overline{\boldsymbol{S}})$ is admissible, and by (73), $\boldsymbol{D} = \boldsymbol{C}\boldsymbol{B}$.

The matrix $\boldsymbol{C}$ can also be expressed by Cayley's formula. Since $\frac{1}{2}(\boldsymbol{E}+\boldsymbol{C})$ is again a triangular matrix, we can set

$$\boldsymbol{C} = 2\boldsymbol{L}^{-1} - \boldsymbol{E},$$

with an indeterminate matrix $\boldsymbol{L}$ of type $\{\kappa\}$. The condition

$$\overline{\boldsymbol{S}}[2\boldsymbol{L}^{-1} - \boldsymbol{E}] = \overline{\boldsymbol{S}}$$

implies that

$$2\overline{\boldsymbol{S}} = \boldsymbol{L}'\overline{\boldsymbol{S}} + \overline{\boldsymbol{S}}\boldsymbol{L}',$$

and hence the matrix

$$\overline{\boldsymbol{S}}^{-1} - \boldsymbol{L}\overline{\boldsymbol{S}}^{-1} = \boldsymbol{A}$$

is alternating. With the decomposition into boxes $\boldsymbol{A} = (\boldsymbol{A}_{pq})$ of type $\{\kappa\}$ it further follows that $\boldsymbol{A}_{pq} = 0$ for $p+q > r$, so that the boxes on and below the anti-diagonal are zero, whereas the remaining boxes are to be chosen real, subject to the condition

$$\boldsymbol{A}_{qp} = -\boldsymbol{A}'_{pq} \quad (p+q \le r).$$

Then

$$\boldsymbol{L} = \boldsymbol{E} - \boldsymbol{A}\overline{\boldsymbol{S}}, \quad \boldsymbol{C} = \frac{\boldsymbol{E}+\boldsymbol{A}\overline{\boldsymbol{S}}}{\boldsymbol{E}-\boldsymbol{A}\overline{\boldsymbol{S}}}, \quad \boldsymbol{A}\overline{\boldsymbol{S}} = \frac{\boldsymbol{C}-\boldsymbol{E}}{\boldsymbol{C}+\boldsymbol{E}}.$$

But now for every triangular matrix $\boldsymbol{M}$ of type $\{\kappa\}$ we have $(\boldsymbol{E}-\boldsymbol{M})^r = 0$, hence the parameter representation

$$\boldsymbol{C} = \boldsymbol{E} + 2\sum_{p=1}^{r-1} (\boldsymbol{A}\overline{\boldsymbol{S}})^p, \quad \boldsymbol{D} = \boldsymbol{C}\boldsymbol{B},$$

and conversely

$$\boldsymbol{A}\overline{\boldsymbol{S}} = -\sum_{p=1}^{r-1} \left(\frac{\boldsymbol{E}-\boldsymbol{C}}{2}\right)^p.$$

It is easy to verify that the number of free elements of $\boldsymbol{A}$ and $\overline{\boldsymbol{H}}$ is precisely the dimension of $H(\boldsymbol{S})$.

As an example choose $r = 3$ and let $j_1 = j$, $j_2 = n-2j$, $j_3 = j$. Then

$$\boldsymbol{S} = \begin{pmatrix} 0 & 0 & \boldsymbol{S}_1 \\ 0 & \boldsymbol{S}_2 & \boldsymbol{S}_{23} \\ \boldsymbol{S}_3 & \boldsymbol{S}_{32} & \boldsymbol{S}_{33} \end{pmatrix}, \quad \overline{\boldsymbol{S}} = \begin{pmatrix} 0 & 0 & \boldsymbol{S}_1 \\ 0 & \boldsymbol{S}_2 & 0 \\ \boldsymbol{S}_3 & 0 & 0 \end{pmatrix}, \quad \boldsymbol{B} = \begin{pmatrix} \boldsymbol{E}_1 & \boldsymbol{S}_3^{-1}\boldsymbol{S}_{32} & \frac{1}{2}\boldsymbol{S}_3^{-1}\boldsymbol{S}_{33} \\ 0 & \boldsymbol{E}_2 & 0 \\ 0 & 0 & \boldsymbol{E}_3 \end{pmatrix},$$

where $\boldsymbol{E}_p$ $(p = 1,2,3)$ denotes the identity matrix with j_p rows. If we define

$$2\boldsymbol{A} = \begin{pmatrix} \boldsymbol{G} & -\boldsymbol{F}' & 0 \\ \boldsymbol{F} & 0 & 0 \\ 0 & 0 & 0 \end{pmatrix} \tag{76}$$

with alternating $\boldsymbol{G}$ with j rows and arbitrary $\boldsymbol{F}$ with $n-2j$ rows and j columns, then

$$2\boldsymbol{A}\overline{\boldsymbol{S}} = \begin{pmatrix} 0 & -\boldsymbol{F}'\boldsymbol{S}_2 & \boldsymbol{G}\boldsymbol{S}_1 \\ 0 & 0 & \boldsymbol{F}\boldsymbol{S}_1 \\ 0 & 0 & 0 \end{pmatrix},$$

$$\boldsymbol{D} = \begin{pmatrix} \boldsymbol{E}_1 & \boldsymbol{S}_3^{-1}\boldsymbol{S}_{32} - \boldsymbol{F}'\boldsymbol{S}_2 & \frac{1}{2}\boldsymbol{S}_3^{-1}\boldsymbol{S}_{33} + (\boldsymbol{G} - \frac{1}{2}\boldsymbol{S}_2[\boldsymbol{F}])\boldsymbol{S}_1 \\ 0 & \boldsymbol{E}_2 & \boldsymbol{F}\boldsymbol{S}_1 \\ 0 & 0 & \boldsymbol{E}_3 \end{pmatrix}$$

and

$$\overline{\boldsymbol{H}} = [\boldsymbol{H}_1, \boldsymbol{H}_2, \boldsymbol{H}_3], \quad \boldsymbol{H}_2\boldsymbol{S}_2^{-1}\boldsymbol{H}_2 = \boldsymbol{S}_2, \quad \boldsymbol{H}_3 = \boldsymbol{H}_1^{-1}[\boldsymbol{S}_1] \tag{77}$$

with positive $\boldsymbol{H}_1$, $\boldsymbol{H}_2$ with j rows and $n-2j$ columns. Since $\boldsymbol{S}_2$ has signature $m-j$, $n-m-j$, $\boldsymbol{H}_2$ depends on $(m-j)(n-m-j)$ parameters. Moreover, the matrices $\boldsymbol{G}$, $\boldsymbol{F}$, $\boldsymbol{H}_1$ respectively contain $\frac{1}{2}j(j-1)$, $j(n-2j)$, $\frac{1}{2}j(j+1)$ parameters, and so indeed

$$(m-j)(n-m-j) + \frac{1}{2}j(j-1) + j(n-2j) + \frac{1}{2}j(j+1) = m(n-m).$$

The preceding formulas hold accordingly for $j_2 = 0$.

It is important that with $\boldsymbol{P}$ and $\boldsymbol{Q}$ also the whole geodesic line through $\boldsymbol{P}$ and $\boldsymbol{Q}$ is contained in the homogeneous space $H(\boldsymbol{S})$. With the notation of Section 8, let $\boldsymbol{P} = \boldsymbol{C}'\boldsymbol{C}$ and $\boldsymbol{Q} = \boldsymbol{T}[\boldsymbol{C}]$. From

$$\boldsymbol{P}\boldsymbol{S}^{-1}\boldsymbol{P}\boldsymbol{S}^{-1} = \boldsymbol{E} = \boldsymbol{Q}\boldsymbol{S}^{-1}\boldsymbol{Q}\boldsymbol{S}^{-1}$$

follows

$$\boldsymbol{P}\boldsymbol{S}^{-1}\boldsymbol{P} = \boldsymbol{Q}\boldsymbol{S}^{-1}\boldsymbol{Q}, \quad \boldsymbol{C}\boldsymbol{S}^{-1}\boldsymbol{C}' = \boldsymbol{T}\boldsymbol{C}\boldsymbol{S}^{-1}\boldsymbol{C}'\boldsymbol{T}.$$

If we set $\boldsymbol{C}\boldsymbol{S}^{-1}\boldsymbol{C}' = (v_{kl})$, $\boldsymbol{T} = [t_1, \ldots, t_n]$, then

$$v_{kl}(t_k t_l - 1) = 0 \quad (k, l = 1, \ldots, n).$$

So for $v_{kl} \neq 0$ we have $t_k t_l = 1$ and then also $t_k^\rho t_l^\rho = 1$ for any real ρ. So in any case

$$v_{kl}(t_k^\rho t_l^\rho - 1) = 0 \quad (k, l = 1, \ldots, n),$$

and with $\boldsymbol{H} = \boldsymbol{C}'\boldsymbol{T}^\rho\boldsymbol{C}$,

$$\boldsymbol{P}\boldsymbol{S}^{-1}\boldsymbol{P} = \boldsymbol{H}\boldsymbol{S}^{-1}\boldsymbol{H}, \quad \boldsymbol{H}\boldsymbol{S}^{-1}\boldsymbol{H}\boldsymbol{S}^{-1} = \boldsymbol{E},$$

which proves the claim.

Now the invariant metric on the homogeneous space shall be expressed via the coordinates $\overline{\boldsymbol{H}}$, $\boldsymbol{A}$. Again, a dot denotes the differential with respect to a curve parameter. By (72),

$$\boldsymbol{H}^{-1}\dot{\boldsymbol{H}} = \boldsymbol{D}^{-1}\dot{\boldsymbol{D}} + \boldsymbol{D}^{-1}\overline{\boldsymbol{H}}^{-1}\boldsymbol{D}\dot{\overline{\boldsymbol{H}}} + \boldsymbol{H}^{-1}\dot{\boldsymbol{D}}'\overline{\boldsymbol{H}}\boldsymbol{D}.$$

If $\sigma(\boldsymbol{M})$ denote the trace of a quadratic matrix $\boldsymbol{M}$, then

$$\begin{aligned}\sigma(\boldsymbol{H}^{-1}\dot{\boldsymbol{H}}\boldsymbol{H}^{-1}\dot{\boldsymbol{H}}) &= \sigma(\overline{\boldsymbol{H}}^{-1}\dot{\overline{\boldsymbol{H}}}\overline{\boldsymbol{H}}^{-1}\dot{\overline{\boldsymbol{H}}}) + 2\sigma(\dot{\boldsymbol{D}}\boldsymbol{D}^{-1}\dot{\boldsymbol{D}}\boldsymbol{D}^{-1}) \\ &\quad + 2\sigma(\overline{\boldsymbol{H}}[\dot{\boldsymbol{D}}\boldsymbol{D}^{-1}]\overline{\boldsymbol{H}}^{-1}) + 4\sigma(\overline{\boldsymbol{H}}^{-1}\dot{\overline{\boldsymbol{H}}}\dot{\boldsymbol{D}}\boldsymbol{D}^{-1}).\end{aligned}$$

Moreover,

$$\dot{C} = ((E - A\overline{S})^{-1}(E + A\overline{S}))^{\cdot} = (E - A\overline{S})^{-1}\dot{A}\overline{S}(C + E),$$
$$\dot{D}D^{-1} = \dot{C}C^{-1} = 2(E - A\overline{S})^{-1}\dot{A}(E + \overline{S}A)^{-1}\overline{S},$$

hence

$$\sigma(H^{-1}\dot{H}H^{-1}\dot{H}) = \sigma(\overline{H}^{-1}\dot{\overline{H}}\overline{H}^{-1}\dot{\overline{H}}) + 8\sigma(\overline{H}[(E - A\overline{S})^{-1}\dot{A}]\overline{H}[(E - A\overline{S})^{-1}]). \tag{78}$$

As the parameters $\overline{H}$ and A are independent of one another, $H(S)$ is the direct product of the spaces of $\overline{H}$ and A. According to (78) we have the corresponding additive decomposition for the metric. From this we can easily obtain an explicit expression for the volume form on $H(S)$. For later use, we only do this for $r = 3$ and use (76) and (77). Let $\{H_1\}$, $\{F\}$, $\{G\}$ denote the Euclidean volume elements in the spaces of H_1, F, G, where in $H_1 = (h_{kl})$ and $G = (g_{kl})$ the quantities h_{kl} $(1 \le k \le l \le j)$ and g_{kl} $(1 \le k < l \le j)$ are taken as independent Cartesian coordinates, and furthermore, in $F = (f_{kl})$, take all the f_{kl} $(k = 1, \dots, n-2j;\ l = 1, \dots, j)$. Moreover, let dv, dv_1, dv_2 denote the volume elements for the metrics $\sigma(H^{-1}\dot{H}H^{-1}\dot{H})$, $\sigma(H_1^{-1}\dot{H}_1H_1^{-1}\dot{H}_1)$, $\sigma(H_2^{-1}\dot{H}_2H_2^{-1}\dot{H}_2)$. By (74),

$$\overline{H}^{-1}\dot{\overline{H}} = -\overline{S}^{-1}\dot{\overline{H}}\overline{H}^{-1}\overline{S},$$

so in the current situation

$$\sigma(\overline{H}^{-1}\dot{\overline{H}}\overline{H}^{-1}\dot{\overline{H}}) = 2\sigma(H_1^{-1}\dot{H}_1H_1^{-1}\dot{H}_1) + \sigma(H_2^{-1}\dot{H}_2H_2^{-1}\dot{H}_2).$$

A simple computation further yields

$$\sigma(\overline{H}[\dot{D}D^{-1}]\overline{H}^{-1}) = \sigma(H_1[\dot{G} + \tfrac{1}{2}\dot{F}'S_2F - \tfrac{1}{2}F'S_2\dot{F}]H_1) + 2\sigma(H_2[\dot{F}]H_1)$$

and finally

$$dv = 2^{\frac{j(j+1)}{4}}dv_1dv_2 2^{\frac{j(j-1)}{2}}|H_1|^{\frac{j-1}{2}}\{G\}2^{j(n-2j)}|H_1|^{\frac{n}{2}-j}|H_2|^{\frac{j}{2}}\{F\}$$

with

$$dv_1 = 2^{\frac{j(j-1)}{4}}|H_1|^{-\frac{j+1}{2}}\{H_1\},$$

so that

$$dv = 2^{j(n-j-\frac{1}{2})}|H_1|^{\frac{n}{2}-j-1}\{H_1\}|H_2|^{\frac{j}{2}}dv_2\{F\}\{G\}. \tag{79}$$

14 Boundary points of the homogeneous space

In the following we assume throughout that $S = (s_{kl})$ is integral. Then the determinant of S is $|S| = (-1)^{n-m}d$ with an integer d. Due to (68), $|H| = d$ for all points in the homogeneous space. If we choose $h = d$, then $H(S)$ lies in P. In particular, $H = \overline{P}(\rho)$, where $\overline{P}(\rho)$ is defined by the inequalities (3) for the normal coordinates. By a Jacobi transformation we obtain

$$H = T[D], \quad T = [t_1, \ldots, t_n], \quad D = (d_{kl}),\ d_{kk} = 1,\ d_{kl} = 0\ (k > l).$$

We will show that all the products $t_k t_{n-k+1}$ $(k = 1, \ldots, n)$ are contained within positive bounds that only depend on n, d and ρ. Such bounds shall henceforth be denoted by $\rho_1, \ldots, \rho_5$.

Let L be the matrix of permutations $x_k \mapsto x_{n-k+1}$ $(k = 1, \ldots, n)$ with only zeros outside the anti-diagonal and only ones on it. Then $T^{-1}[L] = [t_n^{-1}, \ldots, t_1^{-1}]$ and $LC'^{-1}L$ is a triangular matrix whose elements are less than ρ_1. Hence

$$H^{-1}[L] = T^{-1}[L][LC'^{-1}L] \in \overline{P}(\rho_1).$$

At the same time $H \in H(S)$, so that

$$H^{-1}[L][LS] = H. \tag{80}$$

When applying (23) to H and to $H^{-1}[L]$ instead of H, we obtain the following inequalities for the diagonal element h_l of H,

$$\rho_2 t_l > h_l > \rho_3^{-1} \sum_{k=1}^{n} t_k^{-1} s_{kl}^2 \quad (l = 1, \ldots, n).$$

If g is a number in the sequence $1, \ldots, n$, then because of $|S| \neq 0$ there exists one among the $g(n-g+1)$ values s_{kl} $(k = 1, \ldots, g;\ l = 1, \ldots, n-g+1)$ that is different from zero, and then by the integrality of the s_{kl} even $s_{kl}^2 \geq 1$. So for this pair $k = k_g$, $l = l_g$,

$$\rho_2 t_l > \rho_3^{-1} t_k^{-1}.$$

Now, $t_m \leq \rho t_{m+1}$ $(m = 1, \ldots, n-1)$, and by $k \leq g$, $l \leq n-g+1$ it follows that

$$t_g t_{n-g+1} > \rho_4^{-1} \quad (g = 1, \ldots, n).$$

On the other hand,

$$\prod_{g=1}^{n} (t_g t_{n-g+1}) = \prod_{g=1}^{n} t_g^2 = |T|^2 = d^2,$$

hence

$$t_g t_{n-g+1} < \rho_5 \quad (g = 1, \ldots, n),$$

which proves the claim.

Let $H \in H(S)$ and $H \Rightarrow \alpha$. We call α a *boundary point* of the homogeneous space. Then t_k and t_{n-k+1}^{-1} have the same order in the limit, for each $k = 1, \ldots, n$. Hence if $\{\kappa\} = \{\kappa_1, \ldots, \kappa_{r-1}\}$ is the type of α, then $\kappa_p + \kappa_{r-p} = n$ $(p = 0, \ldots, r)$. It follows that the type is symmetric, and by an earlier convention we may assume $r = 2\nu - 1$ to be an odd number by admitting the value 0 for j_ν. The boundary points of $H(S)$ are thus all of symmetric type. Since $H \Rightarrow \alpha$, $H^{-1}[L] = T^{-1}[L][LC'^{-1}L]$ also converges to a boundary point of the same type $\{\kappa\}$. According to (80), Theorem 1 yields a decomposition of the matrix LS into boxes K_{pq} $(p, q = 1, \ldots, r)$ corresponding to the type $\{\kappa\}$, where $K_{pq} = 0$ for $p > q$. Hence $S[x]$ is of the form (70), and we will say that S is of *type* $\{\kappa\}$.

Now introduce normal coordinates of type $\{\kappa\}$. Then we can use the parameter representation from the previous section. We further have to set

$$H_p = \tau_p R_p, \quad \tau_p > 0, \quad |R_p| = 1 \quad (p = 1, \ldots, r). \tag{81}$$

By (75),

$$\tau_p \tau_q R_q = R_p^{-1}[S_p] \quad (p + q = r + 1), \tag{82}$$

hence

$$(\tau_p \tau_q)^{j_p} = |S_p|^2 \quad (p + q = r + 1),$$

so that in the case $j_\nu > 0$, τ_ν is a constant depending only on the type, and for $H \Rightarrow \alpha$, H_ν converges itself to a matrix in $H(S_\nu)$. In the case $j_\nu = 0$, $\tau_\nu = 1$. Since in addition the matrix D converges, this also holds for the matrices

$$C = DB^{-1}, \quad A = \frac{C - E}{C + E}\overline{S}^{-1}.$$

For the boundary point α there are then certain coordinates R_p $(p = 1, \ldots, \nu - 1)$, $A_{pq} = -A'_{qp}$ $(p + q \leq r)$ and $H_\nu \in H(S_\nu)$ $(j_\nu > 0)$, where $R_p > 0$ and $|R_p| = 1$. Conversely, if we are given certain values for these coordinates and for any positive τ_p $(p = 1, \ldots, \nu - 1)$ we define the H_p via (81), (82), then the matrix $H = \overline{H}[D]$ converges for $\tau_p / \tau_{p+1} \to 0$ to a boundary point of type $\{\kappa\}$ defined by these coordinates. Thus all boundary points of the homogeneous space $H(S)$ are completely determined. By including these boundary points we obtain the closed space $\overline{H} = \overline{H}(S)$ from $H(S)$. The set of boundary points of type $\{\kappa\}$ shall be denoted by $H_\kappa = H_\kappa(S)$.

The preceding argument can now be extended to arbitrary ideal points. If more generally $H[U^{-1}] \Rightarrow \alpha$ for a fixed unimodular U and $H \in H(S)$, then α is boundary point of the homogeneous space $H(S[U^{-1}])$ in the usual sense. Moreover, $H \Rightarrow \alpha_U$, that is, α_U is the ideal point of the right coset of U in Γ_κ, which is now considered as the corresponding ideal point for the homogeneous space $H(S)$. Then $S[U^{-1}]$ is of type $\{\kappa\}$, and the coordinates of all ideal points α_U for fixed U are obtained by the above procedure. By including all the ideal points into $\overline{H}(S)$, we obtain a space denoted by $H^* = H^*(S)$.

Finally let $\alpha_U = \beta_V$ with $U \in \Gamma$, $V \in \Gamma$, so that α and β are of the same type $\{\kappa\}$ and $U = WV$, $W \in \Gamma_\kappa$. If we set $S[U^{-1}] = Q_1$, $S[V^{-1}] = Q_2$, then Q_1 and Q_2 are both of type $\{\kappa\}$ and $Q_2 = Q_1[W]$, $\beta = \alpha_W$. If, in analogy to (72), we introduce the coordinates $\overline{H}_1$, A_1 and $\overline{H}_2$, A_2 on $H(Q_1)$ and $H(Q_2)$ by the substitutions

$$H[U^{-1}] = \overline{H}_1[D_1], \quad H[V^{-1}] = \overline{H}_2[D_2], \quad C_1 = D_1 B_1^{-1}, \quad C_2 = D_2 B_2^{-1},$$
$$A_1 = \frac{C_1 - E}{C_1 + E}\overline{Q}_1^{-1}, \quad A_2 = \frac{C_2 - E}{C_2 + E}\overline{Q}_2^{-1}, \quad \overline{Q}_2 = \overline{Q}_1[\overline{W}],$$

then

$$\overline{H}_1[D_1 W] = \overline{H}_2[D_2], \quad \overline{H}_2 = \overline{H}_1[\overline{W}], \quad D_1 W = \overline{W} D_2,$$

where the A_2 can be expressed as a rational fraction in A_1 and vice versa. This determines the coordinate transformation for changing to another coset representative V instead of U.

15 Asymptotes in the homogeneous space

In Section 13 it was shown that with two points $\boldsymbol{P}$ and $\boldsymbol{Q}$ in $H(\boldsymbol{S})$, the geodesic line determined by them is also completely contained in $H(\boldsymbol{S})$. It is clear what is meant by an asymptote in $H(\boldsymbol{S})$, namely an asymptote in the usual sense of which two and hence all points are contained in $H(\boldsymbol{S})$. Let $\{\kappa\}$ denote the corresponding type and $\boldsymbol{U}$ a fixed representative of the corresponding coset C. Then $\boldsymbol{H} = \boldsymbol{T}^{\rho}[\boldsymbol{C}] \Rightarrow \alpha_{\boldsymbol{U}}$ for $\rho \to \infty$, where $\alpha_{\boldsymbol{U}}$ is an ideal point of the coset C and ρ is the arc length. Due to $\boldsymbol{H}[\boldsymbol{U}^{-1}] \Rightarrow \alpha$, $\boldsymbol{S}[\boldsymbol{U}^{-1}]$ is of type $\{\kappa\}$.

First, let $\boldsymbol{U} = \boldsymbol{E}$. Then, as in Section 11,

$$\boldsymbol{C} = (\boldsymbol{C}_{pq}), \quad \boldsymbol{C}_{pq} = 0\ (p > q), \quad \boldsymbol{C}'_{pp}\boldsymbol{C}_{pp} = b_p\boldsymbol{R}_p, \quad b_p > 0, \quad |\boldsymbol{R}_p| = 1,$$
$$\boldsymbol{C}_{pp}^{-1}\boldsymbol{C}_{pq} = \boldsymbol{D}_{pq}, \quad \boldsymbol{T} = [t_1, \dots, t_n].$$

Moreover, let $\theta_1, \dots, \theta_r$ be the distinct elements among the $t_1, \dots, t_n$, in ascending order. On the other hand, if we consider the decomposition (72), then it follows from the uniqueness of this decomposition that

$$\boldsymbol{D} = (\boldsymbol{D}_{pq}), \quad \boldsymbol{H}_p = b_p\theta_p^{\rho}\boldsymbol{R}_p\ (p = 1, \dots, r),$$

so that by (81),

$$\tau_p = b_p\theta_p^{\rho} \quad (p = 1, \dots, r)$$

is proved, where for $j_\nu = 0$ the index $p = \nu$ is excluded. This further implies

$$\theta_p\theta_q = 1 \quad (p + q = r + 1)$$

and in particular $\theta_\nu = 1$ for $j_\nu > 0$. Therefore, the coordinates $\boldsymbol{R}_p$ $(p = 1, \dots, r)$ and $\boldsymbol{A}$ are all constant on the asymptote. These ideas carry over without effort to the case of an arbitrary coset C, by replacing $\boldsymbol{H}$ and $\boldsymbol{S}$ by $\boldsymbol{H}[\boldsymbol{U}^{-1}]$ and $\boldsymbol{S}[\boldsymbol{U}^{-1}]$, respectively.

If $\boldsymbol{S}[\boldsymbol{U}^{-1}]$ is of type $\{\kappa\}$, then for every point of the homogeneous space $H(\boldsymbol{S})$ and for given values θ_p $(p = 1, \dots, \nu - 1)$ there exists precisely one asymptote of the coset C that is completely contained in $H(\boldsymbol{S})$. For variable θ_p, the family of these asymptotes thus depends on precisely $\nu - 1$ parameters. They all end in an ideal point $\alpha_{\boldsymbol{U}}$ in $H^*(\boldsymbol{S})$. Conversely, for any fixed ideal point $\alpha_{\boldsymbol{U}}$ of given coordinates $\boldsymbol{R}_p$ $(p = 1, \dots, \nu - 1)$, $\boldsymbol{A}_{pq}$ $(p + q \le r)$, $\boldsymbol{H}_p$, we can easily determine the family of asymptotes in $H(\boldsymbol{S})$ ending in $\alpha_{\boldsymbol{U}}$, where b_p and θ_p appear as parameters.

16 The group of units

Let $\boldsymbol{Q}_1$ and $\boldsymbol{Q}_2$ both be equivalent to $\boldsymbol{S}$, that is, $\boldsymbol{Q}_1 = \boldsymbol{S}[\boldsymbol{U}^{-1}]$, $\boldsymbol{Q}_2 = \boldsymbol{S}[\boldsymbol{V}^{-1}]$, $\boldsymbol{U} \in \Gamma$, $\boldsymbol{V} \in \Gamma$. Suppose $\boldsymbol{Q}_1$, $\boldsymbol{Q}_2$ are both of type $\{\kappa\}$. If there exists a matrix $\boldsymbol{W} \in \Gamma_\kappa$ with $\boldsymbol{Q}_2 = \boldsymbol{Q}_1[\boldsymbol{W}]$, then $\boldsymbol{Q}_1$ and $\boldsymbol{Q}_2$ are called *equivalent* with respect to Γ_κ. However, it is not necessarily $\boldsymbol{U} = \boldsymbol{W}\boldsymbol{V}$ in this situation.

We show that there exist only finitely many classes for this equivalence relation. For this it is enough to determine for given $\boldsymbol{Q}_1$ the unimodular matrix $\boldsymbol{W}$ in Γ_κ in such a way that the elements of $\boldsymbol{Q}_2 = \boldsymbol{Q}_1[\boldsymbol{W}]$ lie between bounds that only depend on n and d. Let α be any boundary point of type $\{\kappa\}$ in $\overline{H}(\boldsymbol{Q}_1)$ and $\boldsymbol{H}_1 \Rightarrow \alpha$, $\boldsymbol{H}_1 \in H(\boldsymbol{Q}_1)$. Choose $\boldsymbol{W}$ such that $\boldsymbol{H}_1[\boldsymbol{W}]$ is reduced. By the result of Section 6 we may now assume $\boldsymbol{W} \in \Gamma_\kappa$. From (80), with $\boldsymbol{H}_1[\boldsymbol{W}]$, $\boldsymbol{Q}_1[\boldsymbol{W}]$ instead of $\boldsymbol{H}$, $\boldsymbol{S}$, the boundedness of $\boldsymbol{Q}_1[\boldsymbol{W}] = \boldsymbol{Q}_2$ follows by Theorem 1, as claimed. Assume moreover representatives $\boldsymbol{Q}_1, \ldots, \boldsymbol{Q}_g$ of all classes with respect to Γ_κ have been fixed, as are the unimodular matrices $\boldsymbol{U}_l$ with $\boldsymbol{S} = \boldsymbol{Q}_l[\boldsymbol{U}_l]$ for $l = 1, \ldots, g$. If then also $\boldsymbol{Q}_l[\boldsymbol{U}] = \boldsymbol{S}$, then $\boldsymbol{S}[\boldsymbol{U}_l^{-1}\boldsymbol{U}] = \boldsymbol{S}$, hence $\boldsymbol{U} = \boldsymbol{U}_l\boldsymbol{F}$, $\boldsymbol{S}[\boldsymbol{F}] = \boldsymbol{S}$. Conversely, $\boldsymbol{S}[\boldsymbol{F}] = \boldsymbol{S}$ and $\boldsymbol{U} = \boldsymbol{U}_l\boldsymbol{F}$ imply again $\boldsymbol{S}[\boldsymbol{U}_l^{-1}] = \boldsymbol{S}[\boldsymbol{U}^{-1}]$. The intersection of $\Omega(\boldsymbol{S})$ and Γ is called the *group of units* $\Gamma(\boldsymbol{S})$ and its elements are the *units* of $\boldsymbol{S}$. The representation $\boldsymbol{H} \mapsto \boldsymbol{H}[\boldsymbol{F}]$ of $\Gamma(\boldsymbol{S})$ is discontinuous on the homogeneous space $H(\boldsymbol{S})$. Note that $\boldsymbol{F}$ and $-\boldsymbol{F}$ yield the same map. Identification of points in $H(\boldsymbol{S})$ equivalent under $\Gamma(\boldsymbol{S})$ gives rise to the space $H(\boldsymbol{S})/\Gamma(\boldsymbol{S})$, denoted by $J = J(\boldsymbol{S})$. Two points $\boldsymbol{P}$ and $\boldsymbol{Q}$ of H have the same image in J if and only if there is a unit $\boldsymbol{F}$ with $\boldsymbol{Q} = \boldsymbol{P}[\boldsymbol{F}]$. In this case, they will be called *associated*. This definition carries over accordingly to the ideal points of H^*.

Now let $\alpha_{\boldsymbol{V}}$ be an ideal point of $H^*(\boldsymbol{S})$. Let $\{\kappa\}$ be the corresponding type and C the right coset $\Gamma_\kappa \boldsymbol{V}$. Moreover, let $\boldsymbol{S}[\boldsymbol{V}^{-1}]$ be equivalent to $\boldsymbol{Q}_l$ with respect to Γ_κ, that is,

$$\boldsymbol{S}[\boldsymbol{V}^{-1}] = \boldsymbol{Q}_l[\boldsymbol{W}], \quad \boldsymbol{W} \in \Gamma_\kappa, \quad \boldsymbol{U} = \boldsymbol{W}\boldsymbol{V}, \quad \boldsymbol{S}[\boldsymbol{U}^{-1}] = \boldsymbol{Q}_l = \boldsymbol{S}[\boldsymbol{U}_l^{-1}]$$

and

$$\boldsymbol{U} = \boldsymbol{U}_l\boldsymbol{F}, \quad \boldsymbol{F} \in \Gamma(\boldsymbol{S}), \quad \boldsymbol{U}_l\boldsymbol{F} = \boldsymbol{W}\boldsymbol{V}.$$

If C_l is the right coset $\Gamma_\kappa \boldsymbol{U}_l$, then thus $C_l\boldsymbol{F} = C$ for some unit $\boldsymbol{F}$ of $\boldsymbol{S}$. We then call the classes C and C_l *associated*. The number of non-associated classes of ideal points of H^* is thus finite for every type $\{\kappa\}$, namely equal to the number g of classes $\boldsymbol{Q}_1, \ldots, \boldsymbol{Q}_g$. If we set $\alpha = \beta_{\boldsymbol{W}}$, then $\alpha_{\boldsymbol{W}} = \beta_{\boldsymbol{U}_l\boldsymbol{F}}$ is associated to the ideal point $\beta_{\boldsymbol{U}_l}$ from the particular class C_l. Now it remains to determine which points in the same class are associated.

So let $\alpha_{\boldsymbol{U}_l} = \beta_{\boldsymbol{U}_l\boldsymbol{F}}$. Then

$$\boldsymbol{U}_l\boldsymbol{F}\boldsymbol{U}_l^{-1} = \boldsymbol{W} \in \Gamma_\kappa, \quad \alpha = \beta_{\boldsymbol{W}}, \quad \boldsymbol{Q}_l[\boldsymbol{W}] = \boldsymbol{Q}_l,$$

and vice versa. We thus need to study the equivalence of the boundary points of $\overline{H}(\boldsymbol{Q}_l)$ with respect to the intersection $\Gamma(\boldsymbol{Q}_l) \cap \Gamma_\kappa = \Gamma_\kappa(\boldsymbol{Q}_l)$, and firstly this group $\Gamma_\kappa(\boldsymbol{Q}_l)$ itself. We assume that $\boldsymbol{S}$ is already of type $\{\kappa\}$ and replace $\boldsymbol{Q}_l$ by this $\boldsymbol{S}$. Now let $\boldsymbol{W} \in \Gamma_\kappa(\boldsymbol{S})$. Then

$$\overline{\boldsymbol{S}}[\boldsymbol{B}\boldsymbol{W}\boldsymbol{B}^{-1}] = \overline{\boldsymbol{S}}.$$

If we further set

$$\boldsymbol{B}\boldsymbol{W}\boldsymbol{B}^{-1} = \tilde{\boldsymbol{W}}\boldsymbol{C},$$

where $\tilde{\boldsymbol{W}}$ denotes the diagonal part of $\boldsymbol{W}$, then clearly $\boldsymbol{C}$ is a triangular matrix of type $\{\kappa\}$, and from the uniqueness of the generalized Jacobi transformation it again follows that

$$\overline{\boldsymbol{S}}[\tilde{\boldsymbol{W}}] = \overline{\boldsymbol{S}}, \quad \overline{\boldsymbol{S}}[\boldsymbol{C}] = \overline{\boldsymbol{S}}.$$

The first equality means that

$$\boldsymbol{W}'_p\boldsymbol{S}_p\boldsymbol{W}_q = \boldsymbol{S}_p \quad (p + q = r + 1).$$

Hence the matrix $\boldsymbol{W}_p$ for $p = 1,\dots,\nu-1$ is to be determined as unimodular in such a way that

$$\boldsymbol{S}_p^{-1}\boldsymbol{W}'_p\boldsymbol{S}_p = \boldsymbol{W}_q^{-1} \quad (q = r+1-p)$$

is integral, and in the case $j_\nu > 0$, $\boldsymbol{W}_\nu$ has to be chosen in the group of units $\Gamma(\boldsymbol{S}_\nu)$. For the solution $\boldsymbol{C}$ of the second equation we have the parameter representation

$$\boldsymbol{C} = \frac{\boldsymbol{E}+\boldsymbol{A}\overline{\boldsymbol{S}}}{\boldsymbol{E}-\boldsymbol{A}\overline{\boldsymbol{S}}} = \boldsymbol{E} + 2\sum_{p=1}^{r-1}(\boldsymbol{A}\overline{\boldsymbol{S}})^p$$

with alternating $\boldsymbol{A} = (\boldsymbol{A}_{pq})$, $\boldsymbol{A}_{pq} = 0$ for $p+q > r$. Here, for given $\tilde{\boldsymbol{W}}$, $\boldsymbol{A}$ has to be chosen such that the matrix $\boldsymbol{W} = \boldsymbol{B}^{-1}\tilde{\boldsymbol{W}}\boldsymbol{C}\boldsymbol{B}$ is integral.

By (71), $2d\boldsymbol{B}$ is integral. Moreover,

$$\boldsymbol{B}^{-1} = (\boldsymbol{E}-(\boldsymbol{E}-\boldsymbol{B}))^{-1} = \boldsymbol{E} + \sum_{p=1}^{r-1}(\boldsymbol{E}-\boldsymbol{B})^p$$

and hence $(2d)^{r-1}\boldsymbol{B}^{-1}$ is also integral. If now $\boldsymbol{W} = \boldsymbol{W}_0$ is given, then $\tilde{\boldsymbol{W}}_0$ is also integral, and for

$$\boldsymbol{C}_0 = \tilde{\boldsymbol{W}}_0^{-1}\boldsymbol{B}\boldsymbol{W}_0\boldsymbol{B}^{-1},$$

at least $(2d)^r\boldsymbol{C}_0$ is integral. Finally,

$$\boldsymbol{A}_0 = -\sum_{p=1}^{r-1}\left(\frac{\boldsymbol{E}-\boldsymbol{C}_0}{2}\right)^p\overline{\boldsymbol{S}}^{-1}$$

and hence the matrix $t\boldsymbol{A}_0$ is also integral, if we choose

$$t = 2^{r-2}(2d)^{r^2-r+1}.$$

Finally, let

$$f = (2d)^r t^{r-2}$$

and

$$\boldsymbol{A} \equiv \boldsymbol{A}_0 \pmod{f}, \quad \tilde{\boldsymbol{W}} \equiv \tilde{\boldsymbol{W}}_0 \pmod{f}.$$

Then clearly

$$\boldsymbol{C} \equiv \boldsymbol{C}_0 \pmod{(2d)^r}, \quad \boldsymbol{W} = \boldsymbol{B}^{-1}\tilde{\boldsymbol{W}}\boldsymbol{C}\boldsymbol{B} \equiv \boldsymbol{W}_0 \pmod{1},$$

and the matrix $\boldsymbol{W}$ formed with $\boldsymbol{A}$, $\tilde{\boldsymbol{W}}$ belongs to $\Gamma_\kappa(\boldsymbol{S})$. Thus one only has to determine the finitely many classes of $\boldsymbol{A}_0$ and $\tilde{\boldsymbol{W}}_0$ modulo f that are possible for $\boldsymbol{W}_0 \in \Gamma_\kappa(\boldsymbol{S})$. In particular, the choice $\boldsymbol{A} \equiv 0 \pmod{f}$, $\tilde{\boldsymbol{W}} \equiv \boldsymbol{E} \pmod{f}$ always yields an admissible $\boldsymbol{W}$.

Now, if

$$\boldsymbol{A}_1 \equiv 0 \pmod{f}, \quad \boldsymbol{A}_2 \equiv 0 \pmod{f}, \quad \tilde{\boldsymbol{W}}_1 \equiv \boldsymbol{E} \pmod{df}, \quad \tilde{\boldsymbol{W}}_2 \equiv \boldsymbol{E} \pmod{df},$$

then form $\boldsymbol{W} = \boldsymbol{W}_1\boldsymbol{W}_2$. Then

$$\tilde{\boldsymbol{W}} = \tilde{\boldsymbol{W}}_1\tilde{\boldsymbol{W}}_2 \equiv \boldsymbol{E} \pmod{df}, \quad \tilde{\boldsymbol{W}}_1\boldsymbol{C}_1\tilde{\boldsymbol{W}}_2\boldsymbol{C}_2 = \tilde{\boldsymbol{W}}\boldsymbol{C},$$

$$\boldsymbol{C} = \tilde{\boldsymbol{W}}_2^{-1}\boldsymbol{C}_1\tilde{\boldsymbol{W}}_2\boldsymbol{C}_2 = \tilde{\boldsymbol{W}}_2^{-1}\frac{\boldsymbol{E}+\boldsymbol{A}_1\overline{\boldsymbol{S}}}{\boldsymbol{E}-\boldsymbol{A}_1\overline{\boldsymbol{S}}}\tilde{\boldsymbol{W}}_2\frac{\boldsymbol{E}+\boldsymbol{A}_2\overline{\boldsymbol{S}}}{\boldsymbol{E}-\boldsymbol{A}_2\overline{\boldsymbol{S}}} \equiv \boldsymbol{E} + 2(\boldsymbol{A}_1+\boldsymbol{A}_2)\overline{\boldsymbol{S}} \pmod{2df},$$

and hence

$$A = -\sum_{p=1}^{r-1}\left(\frac{E-C}{2}\right)^p \overline{S}^{-1} \equiv 0 \ (\mathrm{mod}\ f).$$

Hence the conditions

$$A \equiv 0 \ (\mathrm{mod}\ f), \quad \tilde{W} \equiv E \ (\mathrm{mod}\ df)$$

define a subgroup of finite index in $\Gamma_\kappa(S)$. Note that the elements of $\Gamma_\kappa(S)$ with $\tilde{W} = E$ form an invariant subgroup, where the factor group is the group of the $\tilde{W}$.

Now we investigate the maps $H^* = H[W]$, $W \in \Gamma_\kappa(S)$. Since we need the parameter representation of H obtained in Section 13 in which the matrices C and A already appeared in a different context, replace C and A by $\overline{C}$ and $\overline{A}$. If then

$$H = \overline{H}[D], \quad DB^{-1} = \overline{C}, \quad H^* = \overline{H}^*[D^*], \quad D^*B^{-1} = C^{-1}$$

and again

$$W = B^{-1}\tilde{W}CB,$$

then we obtain

$$H[W] = \overline{H}[\overline{C}\tilde{W}CB] = \overline{H}^*[D^*]$$

with

$$D^*B^{-1} = C^* = \tilde{W}^{-1}\overline{C}\tilde{W}C, \quad \overline{H}^* = \overline{H}[\tilde{W}]$$

and therefore

$$H_p^* = H_p[W_p] \quad (p = 1, \ldots, r), \tag{83}$$

$$A^* = \frac{C^* - E}{C^* + E}\overline{S}^{-1}, \quad C^* = \tilde{W}^{-1}\frac{E + \overline{AS}}{E - \overline{AS}}\tilde{W}\frac{E + A\overline{S}}{E - A\overline{S}}. \tag{84}$$

We will show that for given $\overline{A}$ and $\tilde{W}$, the matrix $(A_{pq}) = A \equiv 0 \ (\mathrm{mod}\ f)$ can be chosen such that $(A_{pq}^*) = A^*$ is bounded. For by (84), the difference $A_{pq}^* - A_{pq}$ depends only on the A_{kl} with $k + l > p + q$. So we can indeed determine the index sum $p + q$ such that the A_{pq}^* are bounded, as claimed. This remark is useful when determining a fundamental domain for $\Gamma_\kappa(S)$ on the set of boundary points $H_\kappa(S)$. Note that the transformation (83) becomes, via (81), the corresponding one for R_p, whereas (84) holds directly for the coordinates $\overline{A}$ and A^* at a boundary point.

We can now without much difficulty transfer the definition of the normal fundamental domain N_κ for Γ_κ given in Section 7 to the group $\Gamma_\kappa(S)$, by using the results above on the properties of this group. On the other hand, we obtain such a fundamental domain for $\Gamma_\kappa(S)$ directly via reduction theory. For every unimodular U let $R(S, U)$ be the part of $H(S)$ determined by the two conditions $H \in H$, $H[U^{-1}] \in R$. If $R(S, U)$ is not empty, then by definition $S[U^{-1}] = Z$ is reduced and moreover bounded by the afore used argument. Choose all possible $Z_1, \ldots, Z_h$ arising this way for Z, and let $S[V_k^{-1}] = Z_k$ with fixed unimodular V_k. In addition, let $R(S)$ denote the union of the $R(S, V_k)$. For $S[U^{-1}] = Z = Z_k$ it then follows that $V_k^{-1}U = F \in \Gamma(S)$, $U = V_kF$ and vice versa, whereas the map $H \mapsto H[F]$ maps the domain $R(S, U)$ to $R(S, V_k)$. Hence $R(S)$ is a fundamental domain for $\Gamma(S)$ in $H(S)$, that is, a model for $J(S)$.

Every ideal point of the subspace $R(\boldsymbol{S},\boldsymbol{V}_k)$ is of the form $\alpha_{\boldsymbol{V}_k}$, where α is a boundary point of $\overline{R}$. If $\{\kappa\}$ is the type of α, then $\boldsymbol{Z}_k$ is also of type $\{\kappa\}$. Suppose $\boldsymbol{Z}_k$ is equivalent to $\boldsymbol{Q}_l$ with respect to Γ_κ, that is, $\boldsymbol{Q}_l = \boldsymbol{Z}[\boldsymbol{W}_k]$, $\boldsymbol{W}_k \in \Gamma_\kappa$. Then

$$\boldsymbol{V}_k^{-1}\boldsymbol{W}_k\boldsymbol{U}_l = \boldsymbol{F}_k \in \Gamma(\boldsymbol{S})$$

and

$$\alpha_{\boldsymbol{V}_k\boldsymbol{F}_k} = \beta_{\boldsymbol{U}_l} \tag{85}$$

with $\beta = \alpha_{\boldsymbol{W}_k}$. Here, $\beta_{\boldsymbol{U}_l}$ is an ideal point of $H^*(\boldsymbol{S})$ in the coset $C_l = \Gamma_\kappa\boldsymbol{U}_l$. Now, for all the $\boldsymbol{Z}_k$ that are equivalent to $\boldsymbol{Q}_l$ with respect to Γ_κ, form the set of points $\beta_{\boldsymbol{U}_l}$ that are associated via (85) to the ideal points of the respective domains $R(\boldsymbol{S},\boldsymbol{Z}_k)$. These yield a fundamental domain for the intersection of the groups $\Gamma(\boldsymbol{S})$ and $\boldsymbol{U}_l^{-1}\Gamma_\kappa\boldsymbol{U}_l$, so in the case $\boldsymbol{Q}_l = \boldsymbol{S}$ for the group $\Gamma_\kappa(\boldsymbol{S})$ itself. The set of points β is evidently a fundamental domain for $\Gamma_\kappa(\boldsymbol{Q}_l)$ in $H_\kappa(\boldsymbol{Q}_l)$. Thus, all the ideal points of $J = H(\boldsymbol{S})/\Gamma(\boldsymbol{S})$ are determined. For every symmetric type we obtain as many connected domains of ideal points of J as determined by the number of non-associated classes C_l, and this equals the number g of representatives $\boldsymbol{Q}_1,\dots,\boldsymbol{Q}_g$. From the earlier investigations on distances it is evident that ideal points of J from non-associated classes always have infinite distance.

17 Null representations

Let $\boldsymbol{V}$ be unimodular and $\boldsymbol{S}[\boldsymbol{V}]$ of symmetric type $\{\kappa\} = \{\kappa_1,\dots,\kappa_{r-1}\}$ with odd $r = 2\nu - 1 > 1$. We set $\kappa_{r-1} = j$ and with $\nu - 1$, j instead of r, n, we introduce the shortened type $\{\hat{\kappa}\} = \{\kappa_1,\dots,\kappa_{\nu-2}\}$. Here, $2j \le n$ and $2j = n$ only if $\kappa_{\nu-1} = \kappa_\nu$. If $\boldsymbol{u}_1,\dots,\boldsymbol{u}_j$ denote the first j columns of the matrix $\boldsymbol{V}$, then

$$\boldsymbol{u}_k'\boldsymbol{S}\boldsymbol{u}_l = 0 \quad (k = 1,\dots,\kappa_p;\ l = 1,\dots,\kappa_q;\ p+q = r),$$

and in particular this holds for the indices $k, l = 1,\dots,j$.

Conversely, assume now

$$\boldsymbol{u}_k'\boldsymbol{S}\boldsymbol{u}_l = 0 \quad (k = 1,\dots,j;\ l = 1,\dots,j), \tag{86}$$

where the matrix $(\boldsymbol{u}_1 \cdots \boldsymbol{u}_j)$ shall be primitive. We can then complete it to a unimodular matrix $\boldsymbol{U}_0$, and more generally we can obtain any such matrix through

$$\boldsymbol{U} = \boldsymbol{U}_0\begin{pmatrix} \boldsymbol{E} & \boldsymbol{G} \\ 0 & \boldsymbol{B}\end{pmatrix}$$

with unimodular $\boldsymbol{B}$ with $n-j$ rows and integral $\boldsymbol{G}$. According to (86),

$$\boldsymbol{S}[\boldsymbol{U}_0] = \begin{pmatrix} 0 & \boldsymbol{K} \\ \boldsymbol{K}' & * \end{pmatrix}, \quad \boldsymbol{S}[\boldsymbol{U}] = \begin{pmatrix} 0 & \boldsymbol{K}\boldsymbol{B} \\ (\boldsymbol{K}\boldsymbol{B})' & * \end{pmatrix}$$

with integral $\boldsymbol{K}$ with j rows and $n-j$ columns. As is well-known, we may choose the unimodular matrix $\boldsymbol{B}$ such that

$$KB = (0\ M)$$

for a matrix M with j rows with only zeros above the anti-diagonal. But then $S[U]$ is certainly of type $\{\kappa\}$, and even of the symmetric type $\{1, 2, \ldots, j, n-j, n-j+1, \ldots, n-1\}$ contained in it.

Now choose $W \in \Gamma_\kappa$ such that $S[UW^{-1}]$ equals the class representative with respect to Γ_κ, say $S[UW^{-1}] = Q_l = S[U_l^{-1}]$. It follows that $UW^{-1}U_l = F \in \Gamma(S)$ and $U = FU_l^{-1}W$. Let P and P_l denote the two matrices formed by the first j columns of U and U_l^{-1}, respectively. Furthermore, let $\hat{W}$ denote the matrix formed by the first j rows of W. Then $P = FP_l\hat{W}$. More generally, for the given value j, consider all primitive matrices P with j columns that satisfy $S[P] = 0$. Two such null representations P and P^* are called *associated of type* $\{\kappa\}$ if

$$P = FP^*\hat{W}, \quad F \in \Gamma(S), \quad \hat{W} \in \Gamma_\kappa.$$

By the preceding, every null representation P is associated to precisely one null representation P_l $(l = 1, \ldots, g)$ of type $\{\hat{\kappa}\}$. As a consequence, for every shortened type $\{\hat{\kappa}\}$, the classes of associated null representations correspond bijectively to the right cosets $C_l = \Gamma_\kappa U_l$ $(l = 1, \ldots, g)$ and hence also to the distinct connected domains that contain the ideal points of $J(S)$.

The largest value of j for a given S shall be denoted by b. As S has signature m, $n-m$ and we assumed $m \le n-m$, we always have $j \le b \le m$. By Meyer's theorem, either $b = m$ or $\frac{n}{2} - 2 \le b < m$. If moreover $j < b$, then by Witt's theorem we can find $b - j$ columns in addition to the j columns u_k $(k = 1, \ldots, j)$ in (86) such that the matrix $(u_1 \cdots u_b)$ is again primitive and (86) holds with b instead of j. Conversely, we can now obtain the null representation with j columns from that with b columns by deleting the last $b - j$ columns. This again can be interpreted as a property of the ideal points in $J(S)$, but we shall not pursue this further.

18 Finiteness of the volume of the fundamental domain

Since the fundamental domain $R(S)$ consists of the finitely many connected domains $R(S, V_k)$, it is enough to prove the finiteness of $H(S) \cap R$. Moreover, by compactness of $\overline{R}$, it is enough to show finiteness of the volume $H(S) \cap U(\alpha)$, where $U(\alpha) = U$ is a neighborhood of the boundary point α in $\overline{R}$. Let $\{\kappa\}$ be the type of α with $r = 2\nu - 1 > 1$, and let $\kappa_{\nu-1} = j$. We introduce the coordinates H_1, H_2, A with $r = 3$ and $j_1 = j$ that were studied as an example in Section 13. Here, H_2 and A are bounded on U, whereas H_1 is reduced there according to (72). For $H \Rightarrow \alpha$, the matrix H_1 tends to 0. If we set $H_1 = (h_{kl})$, $h_{kk} = h_k$ $(k = 1, \ldots, j)$, then we may assume the inequalities $h_j < 1$ to hold on U.

If we introduce the independent variables F and G according to (76), then these remain bounded on U as well. According to (79) it remains only to prove the convergence of the integral of $|H_1|^{\frac{n}{2}-j-1}\{H_1\}$, taken over all reduced H_1 with $h_j < 1$. For the elements h_{kl} $(1 \le k \le l \le j)$ of H_1 it then holds that

$$-h_k \le 2h_{kl} \le h_k\ (k < l), \quad 0 < h_1 \le h_2 \le \cdots \le h_j,$$

and moreover the expression $h_1 \cdots h_j |\boldsymbol{H}_1|^{-1}$ is bounded on the reduced space R_j. It is thus sufficient to prove the convergence of the integral of

$$(h_1 \cdots h_j)^{\frac{n}{2}-j-1}(h_1^{j-1} h_2^{j-2} \cdots h_{j-1})\mathrm{d}h_1 \cdots \mathrm{d}h_j \quad (0 < h_1 < h_2 < \cdots < h_j < 1).$$

Upon substituting $h_k / h_{k+1} = y_k$ $(k = 1,\dots,j-1)$, $h_j = y_j$, the integrand becomes the product of the j expressions

$$y_l^{\frac{l}{2}(n-l-1)-1} \mathrm{d}y_l \quad (0 < y_l < 1;\ l = 1,\dots,j).$$

But now $\frac{l}{2}(n-l-1) > 0$ except for $l = j = n-1$, and this exceptional case only appears for $j = b = m = n-1 = 1$, that is, for a decomposable binary quadratic form $\boldsymbol{S}[\boldsymbol{x}] = x_2(2s_{12}x_1 + s_{22}x_2)$. This shows the finiteness of the volume of $J(\boldsymbol{S})$ in every other case, whereas in the exceptional case, divergence occurs trivially.

Index

2

Reduction of Quadratic Forms, According to Minkowski and Siegel*

André Weil
Translated by Lizhen Ji

The theory fits in the following schema, of which we will meet other examples below. Let G be a semi-simple Lie group; let Γ be a discrete subgroup of G; classically, it is proposed to find, for Γ in G, a "fundamental domain", that is, a system of representatives in G for G/Γ with nice properties (for example, to be a union of submanifolds embedded in G). If K is a compact subgroup of G, G acts on the homogeneous space $H = G/K$ (notation: we will have G act on H on the right; H is hence the set of left cosets Kx for the subgroup K in G); it is immediate that, if Γ is discrete in G, then Γ, as a subgroup of G, acts on H in a "properly discontinuous" way (which means: for every compact subset X of H, there are only finitely many elements $\gamma \in \Gamma$ such that $X \cap X\gamma \neq \emptyset$). In fact, we always take for K a maximal compact subgroup of G; it results from the theory of semi-simple Lie groups that K is defined in a unique way, up to an inner automorphism G; in other words, H is defined in a unique way up to an isomorphism. Without essentially restricting the generality of the problems proposed to study, it can be assumed that G does not admit a compact invariant subgroup; $H = G/K$ is then, in the sense of Cartan, the "Riemannian symmetric space" associated to G.

From the theoretical point of view, it is convenient to take G to be semi-simple in the strict sense (not only in the infinitesimal sense, but in the global sense), that is, to say its center is reduced to the neutral element (and not only of discrete center). From the point of explicit calculations, it is often convenient to compute with matrices, which leads to infinitesimally simple or semi-simple groups having a discrete center (for example, the special linear group $SL(\mathbb{R}, n)$ over $\mathbb{R}$, whose center is $\pm I_n$ if N is even, where I_n is the unit matrix). Hence there will be many instances of abuse of language, for which we apologize in advance.

In the theory of the reduction of quadratic forms in the classical sense, we start from the group $G_0 = PL_+(\mathbb{R}, n)$ (the connected component of the real projective group in n "homogeneous" variables = the quotient of the linear group $L_+(\mathbb{R}, n)$ in n variables of determinant > 0 by its center = the quotient of the special linear group $SL(\mathbb{R}, n)$ of

* Translated from *Réduction des formes quadratiques, d'aprés Minkowski et Siegel*, Séminaire Henri Cartan, tome 10, no. 1 (1957–1958), exp. no. 1, pp. 1–9, also published in Collected works, vol. 2, pp. 360–366.

n variables of determinant 1 by its center), and the discrete group Γ_0, the image in G_0 of the multiplicative group Γ of matrices of determinant ± 1 with coefficients in $\mathbb{Z}$. A maximum compact subgroup of G_0 is the image K_0 in G_0 of the special orthogonal group $SO(\mathbb{R}, n)$ (orthogonal matrices of determinant 1).

Let P be the space of positive non-degenerate quadratic forms in n variable:

$$F(x) = {}^t x \cdot A \cdot x = \sum_{i,j=1}^{n} a_{ij} x_i x_j, \quad {}^t A = A$$

(a vector x, in matrix notation, will always be denoted by a matrix of n rows and 1 column). We write $A \gg 0$ to express the fact that the symmetric matrix A is the matrix of a positive non-degenerate form. The group $L(\mathbb{R}, n)$ acts on P by the law $(A, X) \to {}^t X \cdot A \cdot X$, where $A \in P$ and X is an element of $L(\mathbb{R}, n)$ written as a matrix. Since any form of P can be written as a sum of n squares (of linear forms in x_i), the group acts transitively on P. Since the element I_n of P (that is, the form "standard" $\sum x_i^2$) is invariant by the orthogonal group $O(\mathbb{R}, n)$, P can be identified with the homogeneous space $L(\mathbb{R}, n)/O(\mathbb{R}, n)$.

In the vector space (of dimension $n(n+1)/2$) of all the quadratic forms in $\mathbb{R}^n$ (or, these forms being expressed by means of the canonical basis of $\mathbb{R}^n$, in the space of the symmetric matrices with n rows and n columns), P is defined by the inequalities $P(x) > 0$ $(x \neq 0)$ and thus forms a convex cone; we immediately check that this is open; its boundary is the set of positive degenerate forms. By passing to the quotient by the equivalence relation whose classes of equivalence are the rays ("half-lines") coming from 0, P determines a convex subset P_0 of a projective space of dimension $n(n+1)/2-1$. By passing to quotient, the projective group $G_0 = PL(\mathbb{R}, n)$ acts on P_0; it results from what precedes that it acts there transitively, and that P_0 is identified with G_0/K_0, the Riemannian symmetric space associated to G_0. Minkowski's theory leads to determination, in P_0, of a fundamental domain for the discrete group Γ_0, which is a convex polyhedron (more exactly, the union of the interior of a convex polyhedron and a suitable part of the boundary).

First, $n = 2$ ("binary forms"); we write

$$F(x) = ax^2 + 2bxy + cy^2;$$

P is the cone determined by $a > 0, ac - b^2 > 0$ (the interior of one sheet of a second degree cone in $\mathbb{R}^3$); P_0 is, in the projective plane, the interior of the conic $ac - b^2 = 0$. As $X \to {}^t X \cdot A \cdot X$ is a representation of the linear group in the space of the symmetric forms A, it follows, by passing to the quotient, that the G_0-actions in P_0 are automorphisms of the ambient projective plane, preserving the conical boundary of P_0; P_0 is taken as "Cayley model" of the hyperbolic plane geometry, G thus induces, on P_0, a subgroup of the group of automorphisms of this geometry; we check without difficulty it is precisely the related subset of the latter group (that is, say the "group of non-Euclidean motions"). The correspondence between the "Cayley model" and the half-plane of Poincaré is obtained as follows: to any form $F \in P$, we determine, on the one hand, the point f which it determines in P_0, and on the other hand, that of the roots z of the equation $az^2 + 2bz + c = 0$ whose imaginary part is > 0. This correspondence $f \rightleftharpoons z$ is a one-to-one correspondence between P_0 and the upper half plane of the variable z. The G_0-actions in this half-plane are obviously those of the real homographic group of determinant > 0 (the group of non-Euclidean motions in the Poincaré model).

To determine a point of P, one can, instead of giving oneself a quadratic form in $\mathbb{R}^n$, give a positive non-degenerate quadratic form F on a vector space E of dimension n over $\mathbb{R}$, and a basis $(e_1,\dots,e_n)$ of this space. If $\langle x,y\rangle$ is a scalar product in E associated to F in the usual way, this data in question determine a quadratic form in $\mathbb{R}^n$ given by

$$F\left(\sum_1^n x_i e_i\right) = \sum_{i,j}\langle e_i, e_j\rangle x_i x_j.$$

Let $A = \|a_{ij}\|$, $a_{ij} = \langle e_i, e_j\rangle$; let $X = \|x_{ij}\|$ be an element of the linear group $L(\mathbb{R}, n)$; by definition, the transform of A by X is $A' = {}^tX\cdot A\cdot X$, which is also written as

$$A' = \|a'_{ij}\|,\ \text{with } a'_{ij} = \langle e'_i, e'_j\rangle,\ e'_i = \sum_{k=1}^n x_{ki}e_k.$$

Saying that X has integer coefficients and is of determinant ± 1 is equivalent to saying that the "lattices" (discrete subgroups of rank n of space E) generated by $(e_1,\dots,e_n)$ and by $(e'_1,\dots,e'_n)$ coincide. The set formed by A and all its transforms by the group Γ is therefore the set of matrices $A' = \langle e'_i, e'_j\rangle$ when we let $(e'_1,\dots,e'_n)$ run over the set of all the systems of generators of the lattice generated by $e_1,\dots,e_n$. Suppose we have chosen, in P_0, a system of representatives M_0 for the equivalence relationship determined by the action of the group Γ_0 on P_0. For the moment, let us agree that the matrix A of a quadratic form in $\mathbb{R}^n$ is "reduced" if the point it determines in P_0 belongs to M_0, and also that a basis $(e_1,\dots,e_n)$ of the space E is "reduced" for a quadratic form F in E if the matrix A of $\langle e_i, e_j\rangle$ is reduced. It follows from the above that, given a form F and a lattice Λ in E, there is at least one system of generators of Λ which is a reduced basis for E, and that two such systems necessarily have the same matrix A, so do not differ from each other except by a transformation of the orthogonal group of F. Choosing a system of representatives M_0 is therefore "almost enough" to declare a law which allows one to associate, with the least possible ambiguity, to any pair formed of a quadratic form F and of a lattice Λ a system of generators of Λ. We will formulate such a law according to Minkowski.

Let $n = 2$ first; we will take for e_1 a vector $\neq 0$ of the lattice Λ whose "length" $F(e_1)^{1/2}$ is the smallest possible, then for e_2 a vector whose length is the smallest possible among those who are not of the form te_1 ($t \in \mathbb{R}$). It is clear that there are only finite numbers of vectors e_1 satisfying the first condition, and that there are at least two; elementary geometric consideration (and obvious) shows that there are exactly two, except in the following cases: (a) the Gauss lattice (integer coordinate points in the plane with the form x^2+y^2); (b) the hexagonal lattice (generated by the vectors $(1,0)$ and $(\frac{1}{2},\frac{\sqrt{3}}{2})$ in the plane with x^2+y^2). It follows that, in all these cases, the various possible choices of e_1 are deduced from each other by a rotation leaving Λ invariant (the rotation angle π in the general case, the angle $m\pi/2$, resp. $m\pi/3$, with m an integer, in case (a), resp. case (b)). As for the second vector e_2, we find no less easily that it is determined in a unique way up to the sign (once we have chosen e_1) except in the case of a lattice generated by two vectors $(1,0)$ and $(\frac{1}{2}, y)$, with $|y| \geq \frac{\sqrt{3}}{2}$, in the plane with x^2+y^2. We will fix the sign indetermination, as much as possible, agreeing to take e_2

such that $\langle e_1, e_2\rangle \geq 0$; actually, when this rule leaves an ambiguity, the lattice Λ admits the line $0e_1$ as axis of symmetry, and the possible choices of e_2 are symmetric to each other with respect to this axis.

A basis (e_1, e_2) of the plane provided with a quadratic form F will be called *reduced* if it possesses the properties mentioned above, with respect to the lattice it generates and the form F. A quadratic form $F = ax^2 + 2bxy + cy^2$ in $\mathbb{R}^2$ will be called *reduced* if the canonical basis of $\mathbb{R}^2$, formed of the vectors $(1,0)$ and $(0,1)$, is reduced with respect to this form. If we write the inequalities

$$F((1,0)) \leq F((0,1)) \leq F((\pm 1,1)),$$

we immediately obtain the conditions $a \leq c, |2b| \leq a$, which are therefore necessary for F to be reduced (with of course $a > 0$ since F must be positive non-degenerate). The subsidiary condition imposed on e_2 here is $b \geq 0$. Thus:

$$0 < a \leq c, \quad 0 \leq 2b \leq a.$$

Conversely, these inequalities lead, first of all, to the fact that F is positive non-degenerate (it's clear), then F is reduced (it's easy to check). They determine in P_0, that is, within the conic $ac - b^2 = 0$, a triangle M_0, having a vertex on the conic. Since, given a lattice and a form, the rule above still allows to construct at least a reduced basis, it follows that M_0 contains a complete system of representatives for the group Γ_0 acting on P_0. Because, for given F and Λ, the various possible choices of the reduced bases all can be deduced from each other by rotation or symmetry which preserves Λ, we conclude that M_0 constitutes such a system of representatives. It follows that P_0 is a union of M_0 and all its transforms by Γ_0 ("tiling" of the non-Euclidean plane by the transforms of the fundamental domain). Naturally, these transforms are not mutually disjoint: this is precisely the case of possible ambiguity in the choice of a reduced basis, or, which amounts to the same, the fact that Γ_0 does not act on P_0 "without fixed point". Detailed analysis of the tiling in question is conventional, offers no difficulty and is of no interest for what follows. Note that if, instead of starting from group Γ, integral matrices of determinant ± 1, we had started from the subgroup Γ^+ of Γ formed by matrices of determinant 1, we would have naturally led to a fundamental domain twice large, corresponding to the notion of "proper equivalence" of quadratic forms (instead of the "proper or improper" equivalence used above); for proper equivalence, a form will be called reduced if it satisfies $0 < a \leq c, |2b| \leq a$; these are the classic conditions of Gauss, corresponding to, in the half-plane of Poincaré, the classical fundamental domain for the modular group (determined by $|z| \geq 1, |R(z)| \leq \frac{1}{2}$).

Let us go to the general case. In a vector space E of dimension n over $\mathbb{R}$, a basis $(e_1, \ldots, e_n)$, generating a lattice Λ, will be called *reduced* with respect to a quadratic form F (positive non-degenerate) if it satisfies following two conditions:

(i) For each i, let E_i be the set of vectors $e \in \Lambda$ such that $(e_1, \ldots, e_{i-1}, e)$ is a part of a system of n generators for Λ; then e_i is a vector of minimal length among those of E_i.

(ii) We have $\langle e_i, e_{i+1}\rangle \geq 0$ for $1 \leq i \leq n-1$.

As above, the condition (ii) serves to get rid of, to the extent of possible, of ambiguous sign inherent in the choice of e_i when F and Λ are given. If we let ourselves be guided by the case $n = 2$, we would have, in (i), taken for E_i the set vectors of Λ which are not in the vector space generated by $e_1,\dots,e_{i-1}$; with this stronger condition, it would not have been true that any lattice has a reduced basis with respect to a given form F, from which it is necessary to modify rather strongly the statement of the later results (it should be noted that in the generalizations of the theory, for example to the fields of algebraic numbers, one can not avoid these more complicated statements). Note that for $e \in E_i$, it is necessary and sufficient that the image of e in the quotient Λ/Λ_{i-1} of Λ by the lattice Λ_{i-1} generated by $e_1,\dots,e_{i-1}$ is a "primitive" element of this quotient (that is, not divisible by an integer > 1). It is the same thing to say that E_i is the set of vectors $\sum_j x_j e_j$, where $x_1,\dots,x_n$ are integers such that the largest common divisor $(x_i,\dots,x_n)$ of $x_i,\dots,x_n$ is equal to 1. As a result, for a positive non-degenerate quadratic form $F(x)=\sum_{ij} a_{ij}x_i x_j$ to be reduced (which means that the canonical basis of $\mathbb{R}^n$ is reduced with respect to this form), it is necessary and sufficient that it satisfies the conditions:

(i) $F(x_1,\dots,x_n) \geq a_{ii}$ for any system of integers $(x_1,\dots,x_n)$ such that $(x_i,\dots,x_n)=1$;

(ii) $a_{i,i+1} \geq 0$ for $1 \leq i \leq n-1$.

It is clear, moreover, that condition (i), joined to $a_{11} > 0$ (resp. $a_{11} \geq 0$) implies F to be positive non-degenerate (or positive).

If we notice that with the notations above we have $e_j \in E_i$ and $e_j \pm e_i \in E_j$ whenever $i < j$, we conclude that (i) implies the inequalities

$$a_{ii} \leq a_{jj} \quad (i<j) \tag{I}$$

and

$$|2a_{ij}| \leq a_{ii} \quad (i<j). \tag{II}$$

In addition, a fundamental theorem, due to Minkowski, states that to every n there corresponds a constant $C_n > 0$ such that (i) also implies the inequality:

$$a_{11}a_{22}\cdots a_{nn} \leq C_n \det(A), \quad A = \|a_{ij}\|. \tag{III}$$

Let M be the subset of P defined by conditions (i) and (ii); let M_0 be its image in P_0. The set M is defined by an infinitely many linear inequalities which are homogeneous with respect to a_{ij}, and it is obvious that it is a "positively homogeneous" convex set. As it is obvious that every lattice Λ has at least one reduced basis with respect to a given form, M_0 contains a system of representatives of P_0 with respect to the group Γ_0. We can give a lot of more precise information on this subject (see below), but they are of little importance from the point of view of applications. The really important results are linked to the introduction of two families of open subsets in P (resp. P_0) that we are going to define now.

For each $t > 1$, let $S(t)$ be a subset of P defined by the inequalities:

$$\begin{aligned} &a_{ii} < t a_{i+1,i+1} \quad (1 \leq i \leq n-1), \\ &|2a_{ij}| < t a_{ii} \quad (1 \leq i < j \leq n), \\ &a_{11}a_{22}\cdots a_{nn} < C_n t^n \det(a_{ij}). \end{aligned} \tag{A}$$

Minkowski's theorem shows that $M \subset S(t)$ for $t > 1$ (this is mainly in this form that we will use it). It is clear that the $S(t)$ form an increasing family in t of open subsets of P, such that P is its union. We will note by $S_0(t)$ the image of $S(t)$ in P_0.

On the other hand, the reduction of the quadratic form F to a sum of squares by the method of Babylonian algebras (the method also known, in the Literature, under the names of "Jacobi's method" and "orthogonalization method of Schmidt", unless it is more appropriate to attribute it some Russian scientists...) allows, in one way and a unique one, to write:

$$F(x) = \sum_{i,j=1}^{n} a_{ij} x_i x_j = \sum_{i=1}^{n} d_i \left(x_i + \sum_{j=i+1}^{n} t_{ij} x_j \right)^2$$

or else, in matrix terms, $A = {}^tT \cdot D \cdot T$, where D is the diagonal matrix having $d_1, \ldots, d_n$ for the diagonal coefficients (and 0 everywhere else), and T is the triangular matrix (in the strict sense, that is, having 1 everywhere in the main diagonal) whose coefficients are the t_{ij} for $i < j$, and the δ_{ij} (1 if $i = j$, 0 otherwise) for $j \leq i$. By induction, it is easy to show that d_i, t_{ij} are rational functions of a_{ij}, with denominators $\neq 0$ in P.

With this, let $S'(u)$, for all $u > 1$, be the set of points of P of the form $A = {}^tT \cdot D \cdot T$, where D is a diagonal matrix with diagonal coefficients $d_1, \ldots, d_n$, and T is a triangular matrix in the strict sense, with coefficients t_{ij} for $i < j$, satisfying the inequalities:

$$0 < d_i < u d_{i+1} \quad (1 \leq i \leq n-1); \quad |t_{ij}| < u \quad (1 \leq i < j \leq n). \tag{B}$$

From what has just been said, it is clear that the $S'(u)$ also form an increasing family of open subsets of P, whose union is P. Trivial calculations make it possible to check that every $S(t)$ is contained in some $S'(u)$, and vice versa. It follows, of course, that M is contained in $S'(u)$ for sufficiently large u. We will denote by $S'_0(u)$ the image of $S'(u)$ in P_0.

We owe to Siegel the following very important result (for the convenience of statement, we will agree, for every subset S of P, and every invertible matrix X, to denote by S^X the transform of S by X, i.e., the set of ${}^tX \cdot A \cdot X$ for $A \in S$):

For every $t > 1$ and an integer $m \neq 0$, the set of matrices X with integer coefficients, of determinant m such that $S(t) \cap S(t)^X \neq \emptyset$, is finite.

It follows naturally that one can say the same for $S'(u)$. Since $M \subset S(t)$, we conclude in particular, taking $m = \pm 1$, that for every $t > 1$, $S(t)$ is contained in the union of M and a finite number of transforms of M by elements of the group Γ.

Finally, it can be verified, for example by explicit calculation, that in P_0 considered as a symmetric Riemannian space G_0/K_0, each of the sets (noncompact) $S_0(t)$ is of finite volume, with respect to the "natural" volume element defined in the space P_0 (the volume element that is naturally invariant by G_0, and that this condition determines it in a unique way up to a scalar factor). In view of the above, this is naturally equivalent to saying that M_0 is of finite volume, or that the homogeneous space G_0/Γ_0 is of finite total volume. Determining the volume of this last space (the invariant volume element in G_0 being explicitly chosen once and for all) is due to Minkowski.

For the sake of memory, we add the following, whose interest is historical and aesthetic but in reality it does not seem to be easy to use. First, the convex set M, defined by the inequalities (i) and (ii) combined with the inequality $a_{11} > 0$, is in fact a convex polyhedral cone; in other words, it suffices to define it in terms of a finite number of inequalities taken from those just mentioned. We saw (without proof, but the demonstration in this case is easy) an example, for $n = 2$; for $n = 3$ and $n = 4$, we know an explicit system of inequalities defining M; nothing of the kind is known for general n. Moreover, M is the closure in P of its interior. Finally, not only P is the union of M and all its transforms M^X by elements X of the group Γ (or, more precisely by transformations of M by a system of representatives in Γ of cosets of the center $\{\pm 1_n\}$ of Γ: because this center acts trivially on P), these transformations form a triangulation or "tiling" of P, in the following sense: their interior are disjoint; the intersection of any two of them is a convex polyhedral cone of smaller dimension: every compact subset has no common points with these transformations except for a finite number of them.

Bibliography

We find comprehensive proofs of the above results in:

SIEGEL (Carl Ludwig). - Einheiten quadratischer Formen, Abh. math. Sem. Hamburg Univ., t. 13, 1940, p. 209–239,

and in its mimeographed courses in Göttingen and the Tata Institute.

The following can also be consulted:

MINKOWSKI (Hermann). - Geometrie der Zahlen. - Leipzig and Berlin, B.G. Teubner, 1910; New York, Chelsea, 1953.

MINKOWSKI (Hermann). - Diskontinuitätsbereich für arithmetische Äquivalenz, J. für reine und angew. Math., t. 129, 1905, p. 220–274; Gesammelte Abhandlungen, Band 2, Berlin, B.G. Teubner, 1911, p. 53–100.

WEYL (Hermann). - Theory of reduction for arithmetical equivalence, I., Trans. Amer. Math. Soc., t. 48, 1940, p. 126–164; II., Trans. Amer. Math. Soc., t. 51, 1942, p. 203–231.

WEYL (Hermann). - Fundamental domains for lattice groups in division algebras, Comment. Math. Helvet., t. 17, 1944/45, p. 283–306; and Selecta Hermann Weyl, Basel and Stuttgart, Birkhäuser, 1956, p. 521–553.

3

Groups of Indefinite Quadratic Forms and Alternating Bilinear Forms*

André Weil
Translated by Lizhen Ji

1 Some general concepts

As before, let G be a semi-simple non-compact Lie group, K a maximal subgroup of G, Γ a discrete subgroup of G. The homogeneous space G/K is the Riemannian symmetric space associated to G.

We are led to consider the following properties of Γ:

(I) $v(G/\Gamma) < +\infty$ (v denotes, of course, the invariant volume, or Haar measure, on G; it is invariant on the right and the left, because G is semi-simple, and descends to G/Γ in an obvious way).

(II) There exists an open subset U in G of finite measure (i.e. $v(U) < +\infty$) such that $U\Gamma = G$ (in other words, the image of U in G/Γ is G/Γ), and $U^{-1}U \cap \Gamma$ is finite (that is, there are only finite number of elements $\gamma \in \Gamma$ such that $U\gamma$ meets U).

(III) G/Γ is compact.

It is clear that (III) $\Rightarrow$ (II) $\Rightarrow$ (I). But, between (II) and (III), it is necessary to insert one more property; to state it, we introduce the following notion. Two groups Γ, Γ' are said to be *commensurable* if $\Gamma \cap \Gamma'$ is of finite index in Γ and in Γ'. We show easily that it is an equivalence relation for subgroups of a group G. It follows that the elements $x \in G$ such that $x\Gamma x^{-1}$ is commensurable to Γ form a group $\tilde{\Gamma}$ (called "the group of transformations" of G/Γ), and that, if Γ and Γ' are commensurable, then the associated "groups of transformation" $\tilde{\Gamma}, \tilde{\Gamma}'$ coincide, which implies in particular that $\Gamma' \subset \tilde{\Gamma}$. Given the above, we are also interested in the property:

(M) There exists an open subset U of finite measure, such that $KU = U$ (U is "saturated" with respect to K, or is the inverse image, under the canonical map of G on G/K, of an open subset of G/K), that $U\Gamma = G$, and that $U^{-1}U \cap x\Gamma$ is finite for every $x \in \tilde{\Gamma}$.

* Translated from *Groupes des formes quadratiques indéfinies et des formes bilinéaires alternées*, Séminaire Henri Cartan, tome 10, no. 1 (1957–1958), exp. no. 2, pp. 1–14, also published in Collected works, vol. 2, pp. 367–377.

(We observe that in the group $\tilde{\Gamma}$, any left coset of Γ is contained in a finite union of right cosets, and vice versa, so that, in the last condition of (M), we can write Γx instead of $x\Gamma$ without any change.)

It is clear that (III) $\Rightarrow$ (M) $\Rightarrow$ (II). If (II) is satisfied, Siegel said that Γ is "of first kind." If (M) is satisfied, we propose to say that Γ is "Minkowskian" in G. The theorems stated in the first exposition essentially say that, in the group $G_0 = PL_+(\mathbb{R}, n)$, the group $\Gamma_0 = PL(\mathbb{Z}, n)$ ("modular group") is Minkowskian; in this case, it is easy to check that the group of transformations $\tilde{\Gamma}_0$ associated to Γ_0 is $PL_+(\mathbb{Q}, n)$; we take for the set U the open subsets $S_0(t), S_0'(u)$ introduced previously; the fact that (M) is verified (and not only (II)) is precisely Siegel's theorem. This same example shows that (M) do not imply (III); on the other hand, it is not known (to speak more cautiously, the speaker does not know) if (I), (II), (M) are really distinct. Outside Fuchsians groups, for which there are geometrical ways of definition (by a "fundamental polygon") and, as the others would say, function theoretical techniques (the universal covering of Riemann surfaces with ramifications), it seems that all known groups satisfying (I) are groups defined arithmetically, and, according to Siegel's work, it can be said that all these groups are Minkowskians.

[N.B. We can give all these arithmetic groups a single definition as follows: Let A be a semi-simple algebra on $\mathbb{Q}$, provided with an anti-involutionary automorphism J; $\mathfrak{M}$ a "module" in A (an additive subgroup of finite type of A, such that $\mathfrak{M}\mathbb{Q} = A$). Let $A_{\mathbb{R}}$ be the extension of A to $\mathbb{R}$, provided with the anti-automorphism that extends J; let G be the group of automorphisms of $A_{\mathbb{R}}$ (equipped with J, that is, the group of automorphisms of the algebra $A_{\mathbb{R}}$ which commute with J); G is semi-simple, and it is even the most common "classical" semi-simple Lie group ("classic" means that G has no factor that is one of the exceptional simple groups). We take for Γ the largest subgroup of G that sends $\mathfrak{M}$ to $\mathfrak{M}$.]

It is easy to show that if (II) is satisfied, Γ is *generated* by the elements of $U \cap U\gamma$, thus *of finite type*. One wonders if the *group of relations* between these generators of Γ are themselves of finite type; it is quite simply shown that this is so, at least, if $KU = U$ [work in G/K, and use the known fact that G/K is simply connected; an exercise recommended to readers].

Finally, it is immediate that, if Γ is Minkowskian, so are all groups commensurable to Γ; on the other hand, "we" do not know if it can happen that two commensurable groups, one is "first kind" and the other is not. It is one of the reasons why it is recommended to always work with (M), rather than with (II), despite the apparent complication of the last part of condition (M).

2 Definite forms

To begin with, let F be an *indefinite non-degenerate* quadratic form in a vector space E of dimension n over $\mathbb{R}$. Let G be the orthogonal group of F (the group of automorphisms of E that leave F invariant). Let E' be the dual of E; any bilinear form on E canonically determines, as we know, a linear map from E to E'; in particular, the bilinear form $F(x, y)$ associated with F, that is $F(x) = F(x, x)$, will determine a linear mapping f from E to E', which is symmetric (i.e., ${}^t f = f$) and of rank n.

Let K be a compact subgroup of G; it leaves invariant, as everyone knows, at least one positive nondegenerate form Φ; let φ be the map from E to E' associated with Φ. One can, by the choice of a suitable basis of E (see *Bourbaki, Alg.*, Chap. IX), put F, Φ in the form $F = \sum_i d_i x_i^2$, $\Phi = \sum_i x_i^2$; the matrix of $\varphi^{-1} f$, for this basis, will be the diagonal matrix with coefficients $d_1, \ldots, d_n$. Any automorphism of E which leaves F and Φ invariant obviously commutes with $\varphi^{-1} f$, or in other words, it leaves invariant the subspaces E_v of E formed respectively of eigenspaces of $\varphi^{-1} f$ with respect to the eigenvalues of $\varphi^{-1} f$, eigenvalues which are none other than the d_i (or rather the distinct elements among them). Let E_+ (respectively E_-) be the direct sum of those of the E_v which are associated with eigenvalues > 0 (resp. < 0). Then K is contained in the product of orthogonal groups K_+, K_- of the forms induced respectively in E_+ and in E_- by F; K_+, K_- are compact, since F is positive (resp. negative) nondegenerate on E_+ (resp. E_-). As a result, for K to be a maximal compact subgroup of G, it is necessary and sufficient that we have $K = K_+ \times K_-$, or in other words that K is the group of automorphisms of E that leave invariant F and a positive nondegenerate form Φ such that $\varphi^{-1} f$ has no eigenvalues other than ± 1, or, which amounts to the same thing, such that $(\varphi^{-1} f)^2 = 1$. It is clear then that the points of G/K (the Riemannian symmetric associated to G) are in one-to-one correspondence with Φ possessing this property; it amounts to the same thing to say that they are in one-to-one correspondence with the pairs (E_+, E_-) of complementary subspaces of E such that F induces on E_+ a positive nondegenerate form and on E_- a negative nondegenerate form, and that E_+ and E_- are orthogonal to each other with resect to F; these are determined reciprocally. Moreover, if (p, q) is the signature of F, E_+ and E_- necessarily have the dimensions p, q. Finally, under the law of inertia, if E_+ is a subspace of E of dimension p on which F induces a nondegenerate positive form, F necessarily induces on the orthogonal E_- of E_+ a negative nondegenerate form, and vice versa. This makes it possible to represent, canonically, G/K as an open subset of a Grassmannian (i.e., the Grassmannian of subspaces of E of dimension p and of the subspaces of E of dimension q); these open subsets can easily be defined by explicit inequalities.

As before, let P be the convex cone of all nondegenerate positive forms on E. Any indefinite non-degenerate form F is associated with, by the above discussion, the set $V(F)$ of $\Phi \in P$ such that $(\varphi^{-1} f)^2 = 1$. It is clear that the forms $\Phi \in V(F)$ have the property $\Phi(x) \geq |F(x)|$ for all $x \in E$. These latter inequalities define a convex subset of P (closed in P), of which $V(F)$ is the boundary (as it immediately results from the possibility of reducing at the same time F and any other form of diagonal shape). We have, for every $x \in E$, $|F(x)| = \inf_{\Phi \in V(f)} \Phi(x)$, from where it follows easily that F is determined uniquely up to the sign by $V(F)$. All this remains true, moreover, if F is "definite", but becomes not interesting, $V(F)$ being then reduced to $\{F\}$ or $\{-F\}$. It is customary to say, by abuse of language, that these $\Phi \in V(F)$ are the "majorants" of F (these are in fact the boundary points of the set of all the majorants).

The group $\mathscr{L}(E, E)$ acts transitively on P, as we have seen, by means of $\Phi \to \Phi \circ X$ (for the map φ from E to E', canonically associated to Φ, this is written as $\varphi \to {}^t X \cdot \varphi \cdot X$; so in particular we can write in matrices after choosing a basis). Of course, $V(F \circ X)$ is the transformed set of $V(F)$ by $\Phi \to \Phi \circ X$; in particular, $V(F)$ is invariant by any element X of the orthogonal group G of F; the way in which G acts on $V(F)$ is the same as that obtained by transporting to $V(F)$, using the one-to-one correspondence between G/K

and $V(F)$ defined above, the action of G on the homogeneous space G/K. Therefore, to study the local properties of the bijection $G/K \to V(F)$, it is enough to study them in the neighborhood of the point $\Phi = \sum x_i^2$, for F given by $F = x_1^2 + \cdots + x_p^2 - x_{p+1}^2 - \cdots - x_n^2$. This is not difficult; it is concluded that $V(F)$ is a real analytic manifold, embedded in P, and that $G/K \to V(F)$ is an isomorphism in the real analytic sense (in particular, it is a differentiable map of rank equal to the dimension pq of G/K).

Hermite introduced the ideas outlined above (which we contented ourselves with Bourbaki flavor) for the arithmetic theory ("reduction") of indefinite quadratic forms. The convex polyhedral cone M ("fundamental domain of Minkowski") being defined in P as stated in the preceding exposition, we say that the indefinite form F is *reduced* if $V(F)$ meets M; for that to make sense, of course we need to be in $\mathbb{R}^n$, since the definition of M is relative to a given basis. We say that a form has *integer coefficients* if it is the case with the matrix of the associated bilinear form. It is easy to see that *there are only a finite number of reduced forms with integer coefficients of a given determinant* ($\neq 0$). It comes down to the same thing to show that all the matrices B, symmetric, with integer coefficients, of a given determinant $b \neq 0$, such that $V(B) \cap S'(u) \neq \emptyset$, is finite, for each given value of u. This comes from the property of $S'(u)$ contained in the following trivial lemma:

LEMMA. There is a unimodular matrix π with integer coefficients, such that, for every $u > 1$, there is a $u' > 1$ for which ${}^t\pi \cdot S'(u)^{-1} \cdot \pi$ (the set of matrices ${}^t\pi \cdot A^{-1} \cdot \pi$, for $A \in S'(u)$) is contained in $S'(u')$.

[We will take for π the matrix of the permutation

$$(1, 2, \ldots, n) \to (n, n-1, \ldots, 1),$$

or in other words, the matrix $(\delta_{i,n+1-j})$; the lemma then results from what, in the triangular group, $T \to T^{-1}$ transforms any compact subset into a compact subset (the property which in no way characterizes the triangular group).]

Given this, $V(B) \cap S'(u)$ is the set of $A \in S'(u)$ such that $(A^{-1}B)^2 = 1$, i.e.:

$$A = BA^{-1}B = {}^t(\pi^{-1}B) \cdot ({}^t\pi \cdot A^{-1} \cdot \pi) \cdot (\pi^{-1}B).$$

Thus the integer matrix $\pi^{-1}B$, of determinant b, transforms a point of $S'(u')$ into a point of $S'(u)$; if u and consequently u' are fixed, there is, according to Siegel's theorem, a finite number of matrices likely to do such a thing. QED

[N.B. We have not assumed that B is indefinite, so the case of positive forms is included; this case is also easy to settle directly.]

The foregoing is mainly used to demonstrate that in the orthogonal group G of an indefinite quadratic form F, non-degenerate, with integer coefficients, the "group of units [arithmetic group]" of F, the intersection of G with the group of matrices of determinant ± 1 over $\mathbb{Z}$, is Minkowskian. For this, let B be the matrix of F; from the above, the matrices B_i equivalent to B (i.e., transformations of B by matrices over $\mathbb{Z}$ of determinant ± 1), such that $V(B_i)$ meets $S(t)$, are finite in number (for a choice, fix once and for all $t > 1$). For each, let M_i be a matrix over $\mathbb{Z}$, of determinant ± 1, transforming B into B_i, i.e., $B_i = {}^tM_i \cdot B \cdot M_i$. Then $A \to {}^tM_i^{-1} \cdot A \cdot M_i^{-1}$ is a bijection between $V(B_i)$ and $V(B)$; let U_i be the open subset of $V(B)$, the image by this bijection of $S(t) \cap V(B_i)$; Let U be the union of U_i. We show that U (more exactly, the inverse image in G of the open subset U of G/K, the image by the bijection previously defined between G/K and $V(B)$) has the properties stated in (M). As for the first, everything comes back obviously to prove that, whatever the indefinite form F, the open set $S(t)\cap V(F)$ is of finite measure with respect to the unique invariant measure (compared to group G) defined in $V(F)$; this is done by explicit majorants. To show that U contains a complete system of representatives with respect to the subgroup (this is the second property to check), let $A\in V(B)$; we can transform A into a point ${}^tM\cdot A\cdot M$ of $S(t)$ by means of a matrix M of determinant ± 1 over $\mathbb{Z}$; this one is then in $V(B')$ with $B'={}^tM\cdot B\cdot M$, which implies, by definition of B_i, that B' is one of the B_i, thus $M_i\cdot M^{-1}$ is a "unit" of B, and also that the transformation ${}^t(M\cdot M_i^{-1})\cdot A\cdot(M\cdot M_i^{-1})$ by $M\cdot M_i^{-1}$ is in U_i, so in U; in other words, A is in the transformation of U by $M_i\cdot M^{-1}$. As for the last point to verify, it results from Siegel's theorem.

As in the reduction of positive forms, we can propose to compute (and not only to give an upper bound of) the volume of G/Γ, the unit of volume being explicitly chosen. For this, we do not get away with it so cheaply. The reader is requested to see the report by Siegel.

On the other hand, in the above, we can replace $S(t)$ by the polyhedral cone of Minkowski; we then obtain a "tiling" of G/K by a "fundamental domain" and its transforms by the group of units of B, this tiling offering to the amateurs aesthetic pleasure about almost the same beauty as that of Minkowski in the cone P positive forms. Note, however, because there are a finite number of B_i, the fundamental domain consists not of one but of several pieces, each of which is a kind of convex polyhedron; allocating to each of these pieces a different color, we obtain, for the whole of the tiling, a very picturesque result (see H. Weyl). As much as "we" can know, this is useless.

Lastly, the foregoing makes it possible to decide in which case G/Γ is compact; obviously, it is necessary and sufficient for this that $S(t)\cap V(B')$, or, what amounts to the same, $S'(u)\cap V(B')$ has compact closure in the cone P for any B' equivalent to B. It is the set of $A\in S'(u)$ such that $B'^{-1}AB'^{-1}A=I_n$; for it to be compact, it is necessary and sufficient that the closure of the cone with the vertex 0 which it generates has no common point, other than 0, with the boundary of the cone P (i.e., does not contain any degenerate form $\neq 0$). But this is the set of $A\in S'(u)$ such that $B'^{-1}AB'^{-1}A=\lambda\cdot I_n$, with $\lambda\geq 0$; if A belongs to the closure of this cone and is degenerate, we will have $B'^{-1}AB'^{-1}A=0$. Expressing that A is in the closure of $S'(u)$ and is $\neq 0$, we can write A in the form

$$\begin{pmatrix}0&0\\0&C\end{pmatrix},$$

with C non-degenerate. From $B'^{-1}AB'^{-1}A=0$, we conclude that B'^{-1} is of the form

$$\begin{pmatrix}\star&\star\\\star&0\end{pmatrix},$$

therefore has at least a diagonal coefficient 0. Therefore, B'^{-1}, so also $B'={}^tB'\cdot B'^{-1}\cdot B'$, and also B, "represent" 0 (which means that, if F is the form of the matrix B, $F(x)=0$ has a rational solution $\neq 0$). The converse follows in the same way. In other words, for G/Γ to be compact, it is necessary and sufficient that B does not represent 0 (which can happen for $n = 3$ and $n = 4$, but a theorem of Meyer's theory states that any indefinite form of five variables with integer coefficients "represents" 0).

3 Alternating forms, Siegel's group

This is the next verse of the song; he sings himself on the same air.

Let F be an alternating non-degenerate bilinear form on E; this requires, of course, that E is of even dimension $2n$. Let f be the mapping from E to E' defined by F. Let G be the group of automorphisms of F; let K be a compact subgroup of G; it leaves invariant a non-degenerate positive form Φ, from which arises a map φ from E to E'. The adjoint, with respect to Φ, of the automorphism $\iota = \varphi^{-1} f$ of E, is $-\iota$; it follows that ι is "semi-simple" (from the matrix point of view, that means that ι can be reduced to the diagonal form, if not over $\mathbb{R}$, in any case over $\mathbb{C}$), with all purely imaginary eigenvalues. If ι has at least two distinct eigenvalues which are not conjugate imaginary of each other, E is decomposed into the direct sum of two subspaces each of which is invariant by any automorphism of E who commutes with ι; it is concluded that, pretty much like as in §2 that then K cannot be maximal. In order for K to be maximal, it is necessary and sufficient that ι has only two distinct eigenvalues, which are imaginary conjugates of each other; by multiplying Φ by a scalar factor > 0, we can assume that these eigenvalues are $\pm i$, which amounts to say that $\iota^2 = -1$. In this case, ι determines a complex structure on E (we define in E the scalar multiplication by complex numbers by means of $(\alpha + i\beta)x = \alpha x + \beta\iota x$); we write E_ι for E provided with this structure; for this complex structure, it is immediate that the binary form with complex values $H = \Phi + iF$ is Hermitian positive non-degenerate. (N.B. Here, and in what follows, we denote by Φ without distinction, by abuse of language, either the quadratic form introduced above, or the associated bilinear form.) Since the elements of K commute with ι, they are automorphisms of E with respect to its complex structure; it is clear then that K is the unitary group determined by the Hermitian form H. It therefore always contains a non-discrete center, formed by multiples $e^{it} \cdot 1$ of the identity automorphism (this, in the sense of the complex structure). In the case of indefinite quadratic forms of signature (p, q), the center of the maximum compact subgroup is non-discrete if $p = 2$ or $q = 2$, and in this case only. It is shown that the existence of such center is necessary and sufficient for the existence of a complex structure on G/K invariant by G; we will check it in the present case.

[N.B. The sufficiency of the condition is justified in general as follows: K acts on G/K, with a fixed point which is the point of G/K which corresponds to itself; it thus acts on the tangent vector space of G/K at this point; in this space, each of the two elements of order 4 of the center of K defines an automorphism of square -1, and allows us to define a complex structure, invariant by K. One can do the same in each point; we thus have an almost complex structure; it remains to show that it is integrable. We can see it for example (according to Ehresmann) noting that, in general, for an almost complex structure, the "lack of integrity" is expressed by a "mixed tensor," the one that gives the coefficients of $\overline{\omega}_\beta\overline{\omega}_\gamma$ in the expression of the differentials $d\omega_\alpha$ of forms ω_α of type (1, 0); expressing that this tensor is invariant by the center of K, we find that it vanishes.]

In the end, we see that G/K has been mapped in one-to-one correspondence with the set of complex structures on E for which F is the imaginary part of a positive nondegenerate Hermitian form $H = \Phi + iF$, and also with the set $V(F)$ of real parts Φ of such forms; as in §2, $V(F)$ is a submanifold of the cone P of the positive nondegenerate forms on E, and $G/K \to V(F)$ is a real analytical bijection from G/K to $V(F)$.

If we are in $\mathbb{R}^{2n}$, and we suppose F is given by a matrix with integer coefficients, we prove, exactly as in §2, that the group of arithmetic units of F is Minkowskian in the group of automorphisms of F. In a subsequent presentation, we will define a more or less explicit open subset U of G/K satisfying condition (M); it will be done, at least, for the properly called "Siegel group" (or "modular group of order n"), which is the group of units if F is given in $\mathbb{R}^{2n}$ by an alternating matrix of *determinant* 1. If "we" have a vice, "we" will even define, in the latter case, a "fundamental domain", which, with its transforms, provides a beautiful tiling of the space G/K.

[N.B. It is known that by a possible choice of $2n$ generators for the subgroup $\mathbb{Z}^{2n}$ of vectors with integer coefficients in $\mathbb{R}^{2n}$, any alternating form with integer coefficients can be written as $\sum_i d_i(x_i y_{n+i} - x_{n+i} y_i)$, where the d_i are integers, the "elementary divisors", each of which is a multiple of the precedent. So there is no need for the theory of reduction, in this case, to show that there is, for a given determinant, only a finite number of non-equivalent forms. Moreover, all these forms are equivalent over $\mathbb{Q}$. But it is easy to see that groups of arithmetic units of two equivalent forms over $\mathbb{Q}$ are always commensurable; this is also true, of course, for quadratic forms; but here we can conclude that groups of all alternating forms with integer coefficients are commensurable with the Siegel group.]

We will now deal with complex structure. For this, we introduce the "complexification" of E, which will be denoted by E_c (for typographical reason, instead of the canonical notation $E_{\mathbb{C}}$; it is, as we know, $E \otimes \mathbb{C}$ with its vector space structure over $\mathbb{C}$; E is considered to be embedded in it in the obvious way). Every automorphism ι of E, of square -1, extends to an analogous automorphism of E_c, which determines a decomposition of E_c in direct sum of the subspaces V_i, V_{-i}, the eigenspaces relative to the eigenvalues i resp. $-i$ of ι; V_i, V_{-i} are subspaces of E_c over $\mathbb{C}$, hence are vector spaces over $\mathbb{C}$, of dimension n; we also have $V_{-i} = \overline{V_i}$, where, according to the common usage, the bar denotes the complex conjugate (defined in E_c in the obvious way). If iE denotes the set of pure imaginary vectors of E_c (image of E by $x \to ix$), $E_c = E \oplus iE$ is a direct sum; if $\mathfrak{R}$ ("real part") is the projection of E_c on E it determines, it is immediate that $\mathfrak{R}$ induces on V_i an isomorphism of the complex structure from V_i to the complex structure of E which is determined by ι, i.e., to that of E_ι. So ι is completely determined by the data of V_i. Conversely, let V_i be a subspace of E_c of dimension n over $\mathbb{C}$; in order that $\mathfrak{R}$ induces on V_i a bijection from V_i to E, it is necessary and sufficient that we have one of the equivalent relations: $V_i \cap iE = \{0\}$, $V_i \cap E = \{0\}$, $V_i \cap \overline{V}_i = \{0\}$; when it is so, V_i allows thus to define on E a complex structure by transporting the structure by means of $\mathfrak{R}$, hence an automorphism ι of E of square -1; if we extend ι to E_c, V_i will be the eigenspace of ι with the eigenvalue i.

Let $\mathfrak{G}$ be the complex Grassmannian of subspaces of E_c of dimension n over $\mathbb{C}$; in $\mathfrak{G}$, let $\mathfrak{J}$ be the set of subspaces whose intersection with E is reduced to $\{0\}$; it is an open subset in $\mathfrak{G}$; according to the above, it is identified with the set of all complex structures on E.

If we have given as before, on $E \times E$, a bilinear alternating nondegenerate form F, it can be extended to a form F_c on $E_c \times E_c$; the same for the extension Φ_c of a symmetric form Φ. If we have, on $E \times E$, $\Phi(x, y) = F(-\iota x, y)$ (which is equivalent to the relation $\iota = \varphi^{-1} f$ written at the beginning of this section), the analogous relation will be true for the extensions of F, Φ, ι to E_c; in particular, on V_i, F_c and Φ_c induce an alternating form F' and a symmetric form Φ' such that we have $\Phi' = -iF'$, which obviously requires $F' = 0$. In other words, V_i is then a *maximal isotropic subspace* of F_c ("isotropic" means that F_c induces 0 on V_i; we then find, for example by the choice of a suitable basis, that V_i, being of dimension n, is *maximal* isotropic because F_c is non-degenerate). These spaces form a submanifold (complex analytic) $\mathfrak{B}$ of the Grassmannian $\mathfrak{G}$; we verify without difficulty that the group of automorphisms of E_c which leave F_c invariant (the "complexification" group G) acts transitively on $\mathfrak{B}$. Conversely, let $V_i \in \mathfrak{J} \cap \mathfrak{B}$; since we have $E_c = V_i \oplus V_{-i}$, we can, for every $z \in E_c$, write $z = u + v$, $u \in V_i$, $v \in V_{-i}$, and we then have $\iota z = iu - iv$; if we also write $z' = u' + v'$, with $u' \in V_i$, $v' \in V_{-i}$, we have $F_c(u, u') = 0$ and $F_c(v, v') = 0$ because V_i and hence $V_{-i} = \overline{V_i}$ are isotropic for F_c; this makes it possible to calculate $F_c(-\iota z, z')$ and to see that this bilinear form is symmetric. We must express furthermore Φ is positive non-degenerate on E, which is equivalent to $F(-\iota x, x) > 0$ whatever $x \neq 0$ in E. Or, the isomorphism of E_ι on V_i, the inverse of the isomorphism of V_i on E_ι induced on V_i by $\mathfrak{R}$, is written as $x \to z = x - i \cdot \iota x$ (verification is immediate); taking account of that V_i, $V_{-i} = \overline{V_i}$ are isotropic for F_c, the preceding inequality is written as $iF_c(z, \overline{z}) < 0$ for any $z \neq 0$ in V_i. It is clear that these V_i satisfying this condition form an open subset Ω in $\mathfrak{B}$; the above implies that this open subset is not empty (since every point of the Riemannian symmetric space G/K determines precisely a point of Ω). It is also clear that $V_i \in \Omega$ implies $V_i \in \mathfrak{J}$; indeed, if not, there would exist $z \neq 0$ in $V_i \cap E$, and we would have $z = \overline{z}$, $iF(z, z) < 0$, which is absurd. From the above, G/K therefore is identified with Ω, which is a manifold (complex analytic) embedded in $\mathfrak{B}$, whence the complex structure of G/K. The group G of the automorphisms of F (i.e. automorphisms of E which leave F invariant) acts on $\mathfrak{B}$ and on Ω in an obvious way, leaving invariant the complex analytic structure, and its way of acting on Ω is that which results from, by transport of structure, its action on G/K. This is satisfying.

Taking advantage of the absence of Dieudonné, we will translate it into matrices, to make the connection with Siegel and to get into a state to calculate when we cannot do otherwise (it still happens sometimes). We take a basis $(e_1, \ldots, e_{2n})$ of E over $\mathbb{R}$ for which the matrix of F is

$$\begin{pmatrix} 0 & I_n \\ -I_n & 0 \end{pmatrix}.$$

Let E', E'' be the subspaces spanned over $\mathbb{R}$, respectively, by $(e_1, \ldots, e_n)$ and $(e_{n+1}, \ldots, e_{2n})$; these are maximal isotropic subspaces of E for F. If $V_i \in \Omega$, and that ι, Φ, etc., have the same meaning as above, Φ will be positive non-degenerate on E, thus on E', and we can choose in E' n orthonormal vectors with respect to Φ; they will be so also in E_ι with respect to the Hermitian form $H = \Phi + iF$ (since E' is isotropic for F); so they will form a basis of E_ι over $\mathbb{C}$, which implies that E' and $\iota E'$ are supplementary in E. Then V_i is transversal to the complexification E'_c of E'; indeed, E'_c is the set of $x' + iy'$, with $x' \in E'$, $y' \in E'$; if such a point is in V_i, we have $y' = -\iota x' \in E' \cap \iota E'$, thus $x' = y' = 0$. Similarly, V_i is transversal to E''_c.

Let us choose in V_i n vectors forming a basis of V_i (over $\mathbb{C}$); write them as "columns" (matrices with $2n$ rows and 1 column) in terms of $(e_1,\dots,e_{2n})$ taken as the basis of E_c over $\mathbb{C}$; this gives a matrix of $2n$ rows and n columns (over $\mathbb{C}$), which we can write $\begin{pmatrix}U\\V\end{pmatrix}$, where U,V are two matrices with n rows and n columns; if we change the basis vectors chosen in V_i, this amounts to multiplying U,V on the right by the same invertible square matrix. Since V_i is transversal to E_c'', the matrix U is of rank n, that is, invertible.

Note that V_i is isotropic for F; this is expressed by the formula:

$$({}^tU\ {}^tV)\cdot\begin{pmatrix}0 & I_n\\ -I_n & 0\end{pmatrix}\cdot\begin{pmatrix}U\\V\end{pmatrix}=0,$$

or in other words ${}^tU\cdot V={}^tV\cdot U$. Similarly, note that $iF_c(z,\overline{z})<0$ for all $z\neq 0$ in V_i; it means $(1/i)({}^t\overline{U}\cdot V-{}^t\overline{V}\cdot U)\gg 0$ (the first member is obviously a Hermitian matrix).

Let $Z=VU^{-1}$ be a matrix that is independent of the basis chosen in V_i. The first of the relations above is written as ${}^tZ=Z$; Z is symmetric. The second is written (by multiplying on the right by U^{-1}, and the left by ${}^t\overline{U}^{-1}$, which does not change the fact that the first member is Hermitian, positive, and non-degenerate) as $(1/i)(Z-{}^t\overline{Z})\gg 0$; in other words, if we write $Z=X+iY$ with X,Y symmetric and real, Y must be positive non-degenerate.

We have thus obtained a bijection of Ω, so ultimately of G/K, with the Siegel space $\mathfrak{S}$, formed by symmetric matrices $Z=X+iY$ over $\mathbb{C}$ such that $Y\gg 0$; $\mathfrak{S}$ can be considered as an open of $\mathbb{C}^N$, with $N=n(n+1)/2$, provided with the complex structure induced by that of $\mathbb{C}^N$.

The action of G on $\mathfrak{S}$ can be written down immediately. Indeed, with the notations above, an element of G will be written in the form of a square matrix

$$\begin{pmatrix}A & B\\ C & D\end{pmatrix},$$

where A,B,C,D are matrices of n rows and n columns over $\mathbb{R}$; this matrix must satisfy the condition

$$\begin{pmatrix}{}^tA & {}^tC\\ {}^tB & {}^tD\end{pmatrix}\cdot\begin{pmatrix}0 & I_n\\ -I_n & 0\end{pmatrix}\cdot\begin{pmatrix}A & B\\ C & D\end{pmatrix}=\begin{pmatrix}0 & I_n\\ -I_n & 0\end{pmatrix}$$

which expresses that it leaves F invariant. This matrix acts on V_i, by the matrices U,V, in terms of the formula

$$\begin{pmatrix}U\\V\end{pmatrix}\to\begin{pmatrix}A & B\\ C & D\end{pmatrix}\cdot\begin{pmatrix}U\\V\end{pmatrix}$$

or in other words:

$$(U,V)\to(AU+BV,CU+DV);$$

it acts on Z by the formula

$$Z\to(C+DZ)\cdot(A+BZ)^{-1}.$$

Finally, we can give the action on $\mathfrak{S}$ of the modular group (the subgroup of G formed of matrices with integer coefficients) an interesting, and even important, interpretation. Since F is given in E, the points of $\mathfrak{S}$ are, after what we have seen, in one-to-one correspondence with the complex structures E, which we can put on E, for which F is an imaginary part of a Hermitian form $H \gg 0$. Suppose given at the same time in E a lattice Λ such that F takes integer values on $\Lambda \times \Lambda$; E/Λ is then a torus of dimension (real) $2n$, on which F determines an integer cohomology class of (real) dimension 2. Any complex structure on E determines a complex torus structure on E/Λ (of complex dimension n); for this to be an Abelian variety, it is necessary and sufficient that there exists on E a Hermitian form $\gg 0$ whose imaginary part takes integer values on $\Lambda \times \Lambda$; and when that is so, there is a *positive divisor* on E/Λ belonging to the class of cohomology determined by this imaginary part; equipped with this class, E/Λ is then called a *polarized Abelian variety*. We thus see that every point of $\mathfrak{S}$ corresponds to a structure of Abelian variety on E/Λ polarized by F; so that for two points to correspond to the same structure, it is necessary and sufficient that they can be deduced from each other by an automorphism of E which leaves the form F and the lattice Λ invariant, thus a member of the discrete group Γ of automorphisms of F which have integer coefficients when we take for basis a system of generators of Λ. In other words, the points of $\mathfrak{S}/\Gamma$ (the quotient of $\mathfrak{S}$ by the equivalence relation defined in $\mathfrak{S}$ by the discrete group Γ) are in one-to-one correspondence with structures of Abelian varieties polarized by F that can be defined on E/Λ. When F is of determinant 1 on Λ (that is, all its elementary divisors on Λ equal to 1), the group Γ that one obtains is the Siegel modular group; the corresponding Abelian varieties are called "principally polarized Abelian varieties" (every Jacobian is such a variety).

Bibliography

Minkowski, Hermann. Geometrie der Zahlen. Leipzig und Berlin, B.G. Teubner, 1910; New York, Chelsea, 1953.

Minkowski, Hermann. Diskontinuitätsbereich für arithmetische Äquivalenz, J. für reine und angew. Math., t. 129, 1905, p. 220–274; Gesammelte Abhandlungen, Band 2, Berlin, B.G. Teubner, 1911, p. 53–100.

Siegel, Carl Ludwig. Einführung in die Theorie der Modulfunktionen n-ten Grades, Math. Annalen, t. 116, 1939, p. 617–657.

Siegel, Carl Ludwig. Einheiten quadratischer Formen, Abh. math. Sem. Hamburg Univ., t. 13, 1940, p. 209–239.

Siegel, Carl Ludwig. Discontinuous groups, Annals of Math., t. 44, 1943, p. 674–689.

Siegel, Carl Ludwig. Symplectic geometry, Amer. J. Math., t. 65, 1943, p. 1–86.

Weyl, Hermann. Theory of reduction for arithmetical equivalence, 1., Trans. Amer. math. Soc., t. 48, 1940, p. 126–164; II., Trans. Amer. math. Soc., t. 51, 1942, p. 203–231.

Weyl, Hermann. Fundamental domains for lattice groups in division algebras, Comment. Math Helvet., t. 17, 1944/45, p. 283–306.

4

Discontinuous Subgroups of Classical Groups*

André Weil

1 Introduction

The object of this course is to prepare the way for a study of certain types of discrete subgroups of the real classical groups and the corresponding quotient spaces. The classical groups will be constructed in a rather special way which actually yields all these groups with only a small number of exceptions. The method consists in taking a semi-simple algebra A over the rationals with an involution σ, extending A to an algebra A_R over the real numbers, and considering the connected component G of the group of automorphisms of A_R which commute with σ. G is, in a natural way, an algebraic matric group, and a subgroup G_2 of matrices in G whose elements are rational integers is a discrete subgroup. Discrete subgroups obtained in this way are to form the main object of study. An illustration of the kind of theorem to be studied is given in §10, where conditions for the compacity of G/G_2 are worked out.

The method of study of G/G_2 involves introducing a second involution on A_R which is positive (definitions in §2), and studying the set $P(A_R)$ of positive symmetric elements of A_R with respect to this involution. It turns out that G/G_2 can be identified with a subset of these positive symmetric elements, and that a special type of set W in $P(A_R)$ (an M-domain) covers the image of G/G_2, in a certain sense, only a finite number of times. Attention can then be transferred to the set W, which is arithmetically defined. The study of the set of W depends on a study of $P(A_R)$ which generalizes classical results of Minkowski on the theory of positive definite quadratic forms and their equivalence under transformation by integral matrices. §2 – §6 of these notes are concerned with this theory. The next two sections give a list of the classical groups which can be obtained as indicated above. In §9 some results are obtained on algebras with involutions, and in §10 these are applied, along with the earlier results, to the construction and study of M-domains.

It may appear that the results obtained in this way will depend on the particular way in which the group G is written as a matric group, since the definition of G_Z cer-

*Lecture notes by Andrew H. Wallace, Summer Quarter, 1958, The University of Chicago.

tainly depends on this. However, the properties which are to be of interest eventually are only those which invariant under commensurability, which can be defined as follows:

Two discontinuous subgroups Γ and Γ' of a group G are said to be commensurable if $\Gamma \cap \Gamma'$ is of finite index in each of them.

Now if the group G is an algebraic matric group over Q, and is represented as a matric group in two different ways, then the two subgroups Γ' and Γ'' of integral matrices in these two representations are commensurable. To prove this let G', G'' be the two representations of G as matric groups and write $x' = I_n + (x'_{ij})$ for an element of G', $x'' = I_m + (x''_{\lambda\mu})$ for an element of G''. The isomorphism between G' and G'' is expressed by equations $x_{ij} = P_{ij}(x''_{\lambda\mu})$, $x''_{\lambda\mu} = Q_{\lambda\mu}(x'_{ij})$ where the P_{ij} and $Q_{\lambda\mu}$ are rational functions and in fact can be taken to be polynomials over Q with zero constant terms. If N is a common denominator for all the coefficients in the P_{ij} and $Q_{\lambda\mu}$, then for $x''_{\lambda\mu} \equiv O(N)$, the corresponding x'_{ij} will be integral. Thus Γ' contains Γ''_N, the subgroup of Γ'' consisting of matrices $\equiv I_m(N)$. Similarly $\Gamma'' \supset \Gamma'_N$. Now Γ''_N is of finite index in Γ'', and so the larger group $\Gamma' \cap \Gamma''$ is of finite index in Γ''; and similarly in Γ'. Thus Γ', Γ'' are commensurable.

The result shows that, as far as properties invariant under commensurability are concerned, no generality is lost by the special method used here of constructing G_Z. In particular it is easy to see that the compacity of G/G_Z discussed in §10 is such a property.

References

For general results in the classical theory of algebras:

M. Deuring, *Algebren* (Ergebrisse der Mathematik, 1935).

A. A. Albert, *Structure of Algebras* (A. M. S. Colloquium publications).

Results used on the classical groups over a field:

A. Weil, A remark on classical groups, forthcoming in the *Journal of Indian Math. Soc.*

2 Positive elements of an algebra

Definitions of positivity and boundedness are presently to be made in terms of trace and norm functions on an algebra. If A is an associative algebra over the real number field R, then the mapping which carries $x \in A$ into ax, for a fixed $a \in A$ is a linear mapping $L(a)$ of A into itself. The correspondence $a \to L(a)$ is the left regular representation of A. In terms of a basis of A, $L(a)$ can be expressed by means of a matrix, whose trace $\mathrm{tr}(a)$ is called the trace of a and whose determinant $N(a)$ is called the norm of a. Clearly $\mathrm{tr}(a)$ and $N(a)$ do not depend on the basis used for A.

An involution on an algebra A over R is an anti-automorphism of period two, that is a linear correspondence $x \to x'$ which is one-one and such that $(xy)' = y'x'$ and $(x')' = x$. The involution is said to be positive if, for all $x \neq 0$ in A, $\mathrm{tr}(xx') > 0$.

For a semi-simple algebra A over R the trace function is invariant under an anti-automorphism, or, what is essentially the same, the trace of the left regular representation (defined above) is equal to that of the similarly constructed right regular representation (starting this time with the mapping $x \to xa$). In particular for the involution $x \to x'$ it follows that $\mathrm{tr}(xy')$ is a symmetric bilinear form. Positivity of the involution means that the derived quadratic form is positive definite.

Lemma 2.1. Every semi-simple algebra A over R has a positive involution.

Proof. A is a direct sum of simple algebras, and the trace of an element in A is the sum of its traces in the component algebras. And so it is enough to prove the result where A is simple, that is to say the algebra of square matrices of some order n over (i) the real numbers, (ii) the complex numbers or (iii) the quaternions.

(i) If x is a matrix of order $n \times n$ with real elements, set x' = transpose of x. The trace here is n times the sum of the diagonal elements of the matrix, and so $\mathrm{tr}(xx') > 0$ for $x \neq 0$.

(ii) If x is a matrix of order $n \times n$ with complex elements, take x' = transpose of the matrix obtained from x by taking the complex conjugates of its elements. Here the trace of an element of the algebra is $2n$ times the sum of the diagonal elements. Hence $\mathrm{tr}(xx') = 2n\times$ sum of squares of moduli of elements of the matrix x, and this is > 0 for $x \neq 0$.

(iii) If x is an $n \times n$ matrix of quaternions, take x' = transpose of the matrix whose elements are the conjugates of those of x. Here the conjugate of the quaternion $\alpha + i\beta + j\gamma + k\delta$ is $\alpha - i\beta - j\gamma - k\delta$. This time the trace of xx' is $4n\times$ sum of squares of moduli of elements of x, where the modulus squared of $\alpha + i\beta + j\gamma + k\delta$ is $\alpha^2 + \beta^2 + \gamma^2 + \delta^2$. Hence $\mathrm{tr}(xx') > 0$ for $x \neq 0$.

This completes the proof of the lemma. □

The above concepts can now be applied to define positivity of an element of an algebra. Let A be a finite dimensional semi-simple algebra over R, furnished with a positive involution, denoted, as above, by the correspondence $x \to x'$. Let a be an element of A such that $a' = a$; it is convenient to call such an element symmetric. Then $\mathrm{tr}(x'ay)$; $x, y \in A$ is a symmetric real-valued bilinear form on A. If the corresponding quadratic form $\mathrm{tr}(x'ax)$ is positive definite, a will be said to be positive.

The concepts just described are now to be applied to semi-simple algebras over R obtained in a special way.

Let A be a finite dimensional simple associative algebra over the field of rationals Q. A will be assumed to have a unit element. By the Wedderburn structure theorem, A can be written as $M_n(k)$, the algebra of all $n \times n$ matrices (for suitable n) with elements in a division algebra k. Let $[k:Q]$, the dimension of k over Q, be d.

In what follows that algebra A_R will also be needed, namely the algebra obtained from A by extending the ground field Q to the field R of all real numbers. Again by the Wedderburn theorems, A_R is a semi-simple algebra and so can be written as a direct sum of matrix algebras over real division algebras, that is to say the real numbers, complex numbers or quaternions. Alternatively A_R can be written as $M_n(k_R)$, the algebra of $n \times n$ matrices over k_R, which last is obtained from k by coefficient extension.

By the Lemma 2.1, k_R can be furnished with a positive involution, which will be fixed in the meantime. This involution must next be extended to A_R. As a notational convention small Greek letters will be used for elements of k and k_R, small Roman letters for those of A and A_R. Then an element $x \in A_R$ is a matrix (ξ_{ij}) of order $n \times n$, with $\xi_{ij} \in k_R$. Denoting by $\xi \to \xi'$ the fixed positive involution on k_R, define the involution $x \to x'$ on A_R by setting the element in the i-th row and j-th column of x' equal to ξ'_{ji}. It must be checked that the operation so defined, which is obviously an involution, is positive. To check this, note first that, when $x = (\xi_{ij})$, $\operatorname{tr}(x) = n \operatorname{tr}(\sum_{i=1}^{n} \xi_{ii})$. (Here the trace on the left is on A_R over R, that on the right is on k_R over R.) It follows that $\operatorname{tr}(xx') = n \operatorname{tr}(\sum_{i,j=1}^{n} \xi_{ij}\xi'_{ij})$ and so is positive except when $x = 0$, since $\xi \to \xi'$ is positive on k_R.

Using the positive involution now constructed on A_R, positivity can be defined for the symmetric elements of this algebra, as already explained. It is convenient to give an alternative description of this notion. Let k_R^n denote the set of n-tuples $(\xi_1,\dots,\xi_n)$ of elements of k_R. Elements of k_R^n will be denoted by underlined Roman letters; for example $(\xi_1,\dots,\xi_n)$ will be written as $\underline{x}$, k_R^n can be thought of in several ways. It is a right k_R module, defining $\underline{x}\alpha = (\xi_1\alpha,\dots,\xi_n\alpha)$. If $\underline{e}_i$ is the n-tuple, $(0,\dots,1,\dots,0)$ with 1 in the i-th place the zero elsewhere, then $\underline{x} = \sum \underline{e}_i\xi_i$. On the other hand, if $[k:Q] = [k_R : R] = d$, then k_R^n can be regarded as a vector space of dimension nd over R. Finally, if $\underline{x}$ is thought of as a matrix of one column, A_R operates on k_R from the left by matrix multiplication; here, if $a = (\alpha_{ij})$ and $\underline{x}$ is as above, $a\underline{x}$ is the matrix of one column with $\sum_{j=1}^{n} \alpha_{ij}\xi_j$ as the i-th element. In this matrix context, it is convenient to define $\underline{x}'$ as the one-rowed matrix $(\xi'_1,\dots,\xi'_n)$. Then the product $\underline{x}'a$ can also be defined, namely as the one-rowed matrix with $\sum_{i=1}^{n} \xi'_i\alpha_{ij}$ in the j-th place.

Now if $\underline{x} = (\xi_1,\dots,\xi_n)$, $\underline{y} = (\eta_1,\dots,\eta_n)$ and $a = (\alpha_{ij})$, the function $A(\underline{x},\underline{y}) = \operatorname{tr}(\underline{x}', a\underline{y})$ is a real bilinear form on k_R^n, regarded as a vector space over R.

Lemma 2.2. The element $a \in A_R$ is symmetric and positive if and only if $A(\underline{x},\underline{y})$ (as just introduced) is symmetric and $A(\underline{x},\underline{x})$ is positive definite.

Proof. If $a' = a$ it is obvious that $A(\underline{x},\underline{y}) = A(\underline{y},\underline{x})$. Conversely, let $A(\underline{x},\underline{y}) = A(\underline{y},\underline{x})$ for all $\underline{x}$, $\underline{y}$ in k_R^n. Then $\operatorname{tr}(\sum \xi'_i\alpha_{ij}\eta_j) = \operatorname{tr}(\sum \eta'_i\alpha_{ij}\xi_j)$ for all η_i and ξ_j. Set $\xi_h = 0 (h \neq i)$, $\xi_i = 1$. Then $\operatorname{tr}(\sum \alpha_{ij}\eta_j) = \operatorname{tr}(\sum \eta'_j\alpha_{ji}) = \operatorname{tr}(\sum \alpha'_{ji}\eta_j)$ (the last equality following from the invariance of the trace under the involution). Finally set $\eta_j = 0$ except for $j = h$ and set $\eta_h = (\alpha_{ih} - \alpha'_{hi})'$. Then

$$0 = \operatorname{tr}\Big(\sum (\alpha_{ij} - \alpha'_{ji})\eta_j\Big) = \operatorname{tr}((\alpha_{ih} - \alpha'_{hi})(\alpha_{ih} - \alpha'_{hi})').$$

And since the involution is positive this implies $\alpha_{ih} = \alpha'_{hi}$, or in other words $a' = a$.

Next suppose a is positive in A_R. Then for $x \in A_R$, $x \neq 0$, $\operatorname{tr}(x'ax) > 0$. If $x = (\xi_{ij})$, this means as was remarked above, $n \operatorname{tr}(\sum_{i,j,h} \xi'_{ji}\alpha_{jh}\xi_{hi}) > 0$, the trace now taken in k_R over R. Take x as the matrix with $\underline{x}$ as its first column and this gives at once $A(\underline{x},\underline{x}) > 0$ for $\underline{x} \neq 0$. Conversely suppose that $A(\underline{x},\underline{x})$ is positive definite. Thus $\operatorname{tr}(\sum_{i,j} \xi'_i\alpha_{ij}\xi_j) > 0$ for each $\underline{x} \neq 0$ in k_R^n. Apply this inequality with $\underline{x}$ replaced by each column of the matrix $x = (\xi_{ij})$ in turn, and add the results. It turns out then that a is positive, as required. □

It is sometimes useful to express properties of $A(\underline{x},\underline{y})$ in terms of $f(\underline{x},\underline{y}) = \underline{x}'a\underline{y}$. The proof of the following is trivial verification:

Lemma 2.3. $f(\underline{x},\underline{y})$ is a k_R-valued function on k_R^n, linear (over k_R) in $\underline{y}$ and anti-linear in $\underline{x}$ (i.e., $f(\underline{x}\alpha,\underline{y}) = \alpha' f(\underline{x},\underline{y})$). If $a' = a$ then $f(\underline{x},\underline{y})' = f(\underline{y},\underline{x})$ and conversely.

Proof. The only non-trivial point is the last. If $f(\underline{y},\underline{x}) = f(\underline{x},\underline{y})'$ then $A(\underline{y},\underline{x}) = A(\underline{x},\underline{y})$ and so $a' = a$ by the previous lemma. □

3 Babylonian reduction theorem

The object of this section is to work out a reduction of a bilinear form $f(\underline{x},\underline{y})$ (in the notation of the last section) to a similar form with a diagonal matrix. The reduction will be carried out by means of a triangular unipotent matrix in A_R, that is to say an $n\times n$ matrix with zeros below the diagonal, each diagonal element equal to $1 \in k_R$ and arbitrary elements of k_R above the diagonal. It is clear that the set of such matrices forms a multiplicative group. The Babylonian reduction can be formulated in two equivalent ways:

Theorem 3.1. Let a be a symmetric positive element of A_R. Then there is a triangular unipotent matrix t such that $t'at$ is a diagonal matrix with positive elements of k_R on the diagonal.

Theorem 3.2. (Alternative form of above) Let $f(\underline{x},\underline{y}) = \underline{x}'a\underline{y}$ have positive symmetric matrix a. Then there exists a simultaneous linear change of the variables $\xi_1,\dots,\xi_n$ and $\eta_1,\dots,\eta_n$ to new sets ξ_i^*, η_i^* by means of a triangular unipotent matrix such that $f(\underline{x},\underline{y}) = \sum \xi_i^{*'}\delta_i\eta_i^*$, where $\delta_i \in k_R$ is positive for each $i = 1,\dots,n$.

Before proceeding to the proof of this theorem it is worth while considering the special case in which $k = Q$, and $k_R = R$, the involution $\xi \to \xi'$ being taken as the identity. A_R reduces to the algebra of matrices of some order $n\times n$ with real elements. The symmetric positive elements of A_R are symmetric matrices (in the usual sense) which are the matrices of positive definite quadratic forms. In this case, in the notation of the last section $A(\underline{x},\underline{y}) = f(\underline{x},\underline{y})$ is a real valued bilinear form on an n-dimensional vector space over R, and Theorem 3.1 (or 3.2) becomes the classical reduction theorem for such forms. In fact the following proof is a direct generalization of the usual proof for the classical case.

Proof of Theorem 3.1. Write $f(\underline{x},\underline{y}) = x'ay = \sum_{i,j=1}^n \xi_i'\alpha_{ij}\eta_j$. First separate out the terms containing ξ_1' and η_1:

$$\begin{aligned} f(\underline{x},\underline{y}) =& \xi_1'\alpha_{11}\eta_1 + \xi_1'\alpha_{12}\eta_2 + \dots + \xi_1'\alpha_{1n}\eta_n \\ &+ \xi_2'\alpha_{21}\eta_1 + \dots + \xi_n'\alpha_{n1}\eta_1 \\ &+ \text{terms independent of } \xi_1' \text{ and } \eta_1. \end{aligned} \tag{1}$$

The idea is to show that the terms in (1) containing ξ_1' and η_1 can all be obtained in the expansion of the following:

$$\begin{aligned} f_0(\underline{x},\underline{y}) =& (\xi_1 + \alpha_{11}^{-1}\alpha_{12}\xi_2 + \alpha_{11}^{-1}\alpha_{13}\xi_3 + \dots + \alpha_{11}^{-1}\alpha_{1n}\xi_n)'\alpha_{11} \\ &(\eta_1 + \alpha_{11}^{-1}\alpha_{12}\eta_2 + \dots + \alpha_{11}^{-1}\alpha_{1n}\eta_n). \end{aligned}$$

In this expression α_{11}^{-1} certainly exists. For in a finite-dimensional algebra over a field, the existence of an inverse is equivalent to not being a divisor of zero; and if α_{11} were a divisor of zero, say $\alpha_{11}\eta = 0$, $\eta \neq 0$, then, taking $\underline{x} = \underline{y} = (\eta, 0, \ldots, 0)$, it would follow that $f(\underline{x}, \underline{x}) = 0$, and so $A(\underline{x}, \underline{x}) = \operatorname{tr} f(\underline{x}, \underline{x}) = 0$, with $\underline{x} \neq \underline{0}$, contrary to the positivity of a. Now expanding $f_0(\underline{x}, \underline{y})$, noting that the symmetry of a implies $\alpha'_{ij} = \alpha_{ji}$ and also that $(\alpha'_{11})^{-1} = (\alpha_{11}^{-1})' = \alpha_{11}^{-1}$:

$$\begin{aligned} f_0(\underline{x}, \underline{y}) = &\xi'_1 \alpha_{11}\eta_1 + \xi'_1 \alpha_{12}\eta_2 + \cdots + \xi'_1 \alpha_{1n}\eta_n \\ &+ \xi'_2 \alpha_{21}\eta_1 + \cdots + \xi'_n \alpha_{n1}\eta_1 \\ &+ \text{terms independent of } \xi'_1 \text{ and } \eta_1. \end{aligned} \tag{2}$$

Subtracting (2) from (1) it turns out that $f(\underline{x}, \underline{y}) - f_0(\underline{x}, \underline{y})$ is of the form $\sum_{i,j=2}^{n} \xi'_i \beta_{ij} \eta_j$. Then write $\xi_1^* = \xi_1 + \alpha_{11}^{-1}\alpha_{12}\xi_2 + \cdots + \alpha_{11}^{-1}\alpha_{1n}\xi_n$, and a similar expression for η_1^*, so that $f(\underline{x}, \underline{y}) = \xi_1^{*'} \alpha_{11}\eta_1^* + \sum_{i,j=2}^{n} \xi'_i \beta_{ij}\eta_j$. The theorem will now be completed by induction. It is clearly true if $n = 1$, and will be assume true for $n-1$. It must be checked that the matrix $(\beta_{ij})(i, j = 2, \ldots, n)$ is a positive symmetric element of $M_{n-1}(k_R)$. These conditions follow immediately from the corresponding conditions for a. Then the induction hypothesis shows that a linear change of $\xi_2, \ldots, \xi_n$ and $\eta_2, \ldots, \eta_n$ can be made by means of a unipotent triangular matrix t_0 to new variables $\xi_2^*, \ldots, \xi_n^*$ and $\eta_2^*, \ldots, \eta_n^*$ such that

$$\sum_{i,j=2}^{n} \xi'_i \beta_{ij} \eta_j = \sum_{i=2}^{n} \xi_i^{*'} \delta_i \eta_i^*$$

where the δ_i are positive. Form the matrix t by adding the row $1, \alpha_{11}^{-1}\alpha_{12}, \ldots, \alpha_{11}^{-1}\alpha_{1n}$ to the top of t_0 and zeros down the left hand side. Then t transforms the ξ_i and η_j to the ξ_i^* and $\eta_j^* (i, j = 1, \ldots, n)$ and

$$f(\underline{x}, \underline{y}) = \sum_{i=1}^{n} \xi_i^{*'} \delta_i \eta_i^*$$

where all the δ_i are positive. To prove the last point it is only necessary to note that $\delta_1 = \alpha_{11}$. That this element is positive, can be seen by setting $\underline{x} = \underline{y} = (\xi, 0, \ldots, 0)$ for any ξ. Then $A(\underline{x}, \underline{x}) = \operatorname{tr} f(\underline{x}, \underline{y}) = \operatorname{tr}(\xi' \alpha_{11} \xi)$, and the required result follows from the positivity of $A(\underline{x}, \underline{x})$ for $x \neq \underline{0}$. The proof of the theorem is thus completed. □

4 The Minkowski-Siegel theorem

The object of the present section is to formulate a generalization of the Minkowski theorem on the reduction of positive definite quadratic forms to a special canonical form by means of a linear substitution of bounded determinant. Some preliminary discussion of the proof will also be given, the details being contained in the next section.

Preliminary to the statement of the theorem, some definitions must be given, and in particular the notion of integral elements in k (and k_R) must be explained. The last point is dealt with as follows:

An *order* ϑ in k is a finitely generated additive (Abelian) subgroup of k which generates k over Q and is at the same time a subring of k. It is convenient to assume that ϑ

contains the identity of k. An order is to be selected arbitrarily in k and fixed once and for all. A convenient method is to pick a basis $\omega_1,\dots,\omega_d$ of k over Q with $\omega_1=1$, say. If $\omega_\lambda\omega_\mu=\sum_{\nu=1}^{d}c_{\lambda\mu\nu}\omega_\nu$ then the structural constants $c_{\lambda\mu\nu}$ of k are rational numbers, which can be arranged over a common denominator N, say. Then replacing each ω_λ by $N\omega_\lambda$ (an admissible change of basis) the new structure constants become integers. Assume that this change has been made and let ϑ be the additive Abelian group generated by $\omega_1,\dots,\omega_d$ over Z. Since the structure constants are integers, ϑ is a subring of k; and it clearly generates k over Q. Thus ϑ is an order as defined above. Note that with this construction, the relations between ϑ, k and k_R are simply expressed, these three all being generated by $\omega_1,\dots,\omega_d$, respectively over Z, Q and R.

An element of k, or k_R, will be called *integral* if it belongs to ϑ. It will also be convenient to call a matrix $a\in A_R$ integral if all its elements are integrals in k_R.

The theorem stated below is concerned with positive symmetric elements of k_R and A_R. Let $P(k_R)$ and $P(A_R)$ be, respectively, the sets of all positive symmetric elements of k_R and A_R. In terms of the basis $\omega_1,\dots,\omega_d$ of k_R, which, once chosen, can be assumed fixed for the rest of the discussion, the elements of this algebra have coordinates, in terms of which k_R can be given the topology of a Euclidean space. $P(k_R)$ can then be given the induced topology. In a similar manner $P(A_R)$ becomes a topological space.

The triangular unipotent matrices which appeared in Theorem 3.1 form a set T, which can also be topologized in the manner just described.

In terms of these concepts, a *Siegel domain* in $P(A_R)$ can now be defined. Let K and K' be compact sets in T and $P(k_R)$, respectively, and let C be a constant. Then a Siegel domain $\mathscr{S}$ (depending on K, K', C) is the set of all $a\in P(A_R)$ which can be written in the form $a=t'dt$, where t is unipotent triangular and lies in K, d is a diagonal matrix, diag $\{\delta_1,\dots,\delta_n\}$ such that $\frac{1}{\mathrm{tr}\delta_i}\delta_i\in K'$ for each i, and $0<\mathrm{tr}\delta_i\le C\mathrm{tr}\delta_{i+1}$ for $1\le i\le n-1$.

The statement of the Minkowski-Siegel theorem is:

Theorem 4.1. There is a constant C and a Siegel domain $\mathscr{S}$ such that, for every $a\in P(A_R)$, there exists an integral matrix b such that $0<|N(b)|\le C$ and $b'ab\in\mathscr{S}$.

There are a few remarks to be made about this theorem before the proof is carried out. In the first place it should be noted that C and $\mathscr{S}$ are independent of the matrix a, and depend only on the initial data, namely k, ϑ and the dimension n.

Secondly, the theorem just stated reduces to the classical case if the following special choice are made. Namely take the algebra k to be rational field Q, and the involution $\xi\to\xi'$ to be the identity, and the order ϑ to be the ring of integers Z. The δ_i are in this case positive real numbers, and so the condition $\frac{1}{\mathrm{tr}\delta_1}\delta_i\in K'$ in the definition of the Siegel domain becomes vacuous. $N(b)$ in this case becomes simply the determinant of b, and it turns out that the condition $0<|N(b)|\le 0$ can here be replaced by $\det b=\pm1$.

The classical case can thus be formulated in full as follows: A Siegel domain $\mathscr{S}$ in the space of positive definite real symmetric matrices of order $n\times n$ is the set of matrices of the form $t'dt$, where t is unipotent triangular and d is a diagonal matrix diag$\{\delta_1,\dots,\delta_n\}$, with the δ_i real and positive, and absolute values of the elements of t and the δ_i/δ_{i+1} being all bounded by a preassigned constant C. The domain $\mathscr{S}$ thus depends on C, a point where can be emphasized by writing it as $\mathscr{S}_C$. Theorem 4.1 now takes the form:

There is a Siegel domain $\mathscr{S}$ in the space of positive definite symmetric matrices over R such that, for any a in this space there exists an integral matrix b with $\det b = \pm 1$ such that $b'ab \in \mathscr{S}$. (Here $b' = {}^t b$, the transpose of b.)

A note on the proof of this special case of Theorem 4.1 will be given later.

In the course of the proof of Theorem 4.1, and also in subsequent results, large numbers of absolute constants turn up, that is constants which, like the C in the statement of the theorem, depend only on the initial data of the problem. To avoid having to write these constants explicitly, the following notation will sometimes be used. Two variables X_1 and X_2, taking only positive values, will be said to satisfy the relation $X_1 \prec X_2$ if there is a constant C such that $X_1 \le CX_2$. If $X_1 \prec X_2$ and $X_2 \prec X_1$, X_1 and X_2 will be said to have the same order of magnitude, written $X_1 \asymp X_2$. For example, the inequalities appearing in the definition of the Siegel domain can be written as $\delta_i \prec \delta_{i+1}$ for each i.

A final remark must be made about the norm $N(b)$ which appears in the statement of Theorem 4.1. Instead of taking it as the determinant of the regular representation of A_R, it will be taken as the determinant of the representation obtained by operations on k_R^n, b being mapped on the linear transformation $\underline{x} \to b\underline{x}$. This norm is usually called the *reduced norm*. Its use makes no difference to the statement of the theorem, since the ordinary norm is a power of the reduced norm, but simplifies some of the notation.

The proof of Theorem 4.1 will be preceded by some lemmas. The first of these is Minkowski's theorem on the minimum value assumed by a quadratic form at the points of a lattice.

Lemma 4.2. Let $f = \sum_{i,j=1}^m a_{ij}x_ix_j$ be a positive definite quadratic form in the m real variables x_i, (a_{ij}) being a real symmetric matrix. Then the minimum of f, taken over all integral values of the x_i (not all zero) is $\le C[\det(a_{ij})]^{\frac{1}{m}}$ where C is a numerical constant, independent of the a_{ij}.

Proof. Let L be the lattice of points with integral coordinates in m-dimensional space R^m and let ϕ be the natural mapping of R^m on R^m/L. ϕ induces a measure on R^m/L in such a way that if $X \subset R^m$ is mapped in a one-one manner by ϕ, then the measure of $\phi(X)$ is equal to the volume of X in R^m in the usual sense. In particular $\phi(R^m/L)$ has measure 1. It follows that any set $X \subset R^m$ mapped in a one-one manner by ϕ must have volume less than 1. This means that if the volume of X is greater than 1, X must contain at least two points, $x, x' \in R^m$ such that $x - x' \in L$, $x \neq x'$ (addition of points means vector addition). Now take X to be convex and symmetric in the origin. The point $x' \in X$ may then be replaced by $-x'$. Thus if the volume of X is > 1, there are two points x, x' of X such that $x + x' \in L$. But X is convex and so $\frac{1}{2}(x + x') \in X$. In other words X contains a point $x'' = x + x'$ such that $2x'' \in L$. Changing the scale, it follows that if X is symmetric in the origin, convex, and of volume $> 2^m$, then X contains at least one point of L different from the origin.

In particular, let X be the set of points defined by

$$\sum_{i,j=1}^{m} a_{ij}x_ix_j \le C[\det(a_{ij})]^{\frac{1}{m}}$$

where the constant C is to be independent of the a_{ij}. It is easy to see that the volume of X is independent of the a_{ij}, and by suitable choice of C can certainly be made $>$

2^m. Fixing C in this way, X must contain a point of L other than the origin. Thus the minimum value of $f = \sum a_{ij} x_i x_j$ for integral values of the x_i (not all zero) is certainly less than or equal to $C[\det(a_{ij})]^{\frac{1}{m}}$, as was to be proved. □

The following pair of lemmas give inequalities between norms and traces in k_R.

Lemma 4.3. If δ is a positive symmetric element of k_R, then $\frac{1}{d}\operatorname{tr}(\delta) \geq N(\delta)^{\frac{1}{d}}$.

Proof. By hypothesis $\operatorname{tr}(\zeta'\delta\zeta)$ is a positive definite quadratic form over R. So also is $\operatorname{tr}(\zeta'\zeta)$, by the positivity of the involution. It follows that a basis $\alpha_1, \ldots, \alpha_d$ can be chosen in k_R over R so that, if $\zeta = \sum_{\lambda=1}^d u_\lambda \alpha_\lambda$, then $\operatorname{tr}(\zeta'\zeta) = \sum_{\lambda=1}^d u_\lambda^2$, $\operatorname{tr}(\zeta'\delta\zeta) = \sum_{\lambda=1}^d c_\lambda u_\lambda^2$, where the c_λ are all positive real numbers. Assume $c_1 \leq c_2 \leq \cdots \leq c_d$.

The next step is to find the matrix of the linear mapping on k_R (in terms of the basis $\alpha_1, \ldots, \alpha_d$) corresponding to left multiplication by δ. For any λ, $\operatorname{tr}(\alpha'_\lambda \delta \alpha_\lambda) = c_\lambda$ (this corresponds to taking $u_\lambda = 1$ and the other coordinates $= 0$). Similarly $\operatorname{tr}(\alpha'_\lambda \alpha_\lambda) = 1$. Hence $\operatorname{tr}(\alpha'_\lambda \delta \alpha_\lambda) = c_\lambda \operatorname{tr}(\alpha'_\lambda \alpha_\lambda) = \operatorname{tr}(\alpha'_\lambda c_\lambda \alpha_\lambda)$. On the other hand for $\mu \neq \lambda$, the diagonal form of $\operatorname{tr}(\zeta'\delta\zeta)$ and $\operatorname{tr}(\zeta'\zeta)$ gives $\operatorname{tr}(\alpha'_\mu \delta \alpha_\lambda) = \operatorname{tr}(\alpha'_\mu c_\lambda \alpha_\lambda) = 0$, and so for all μ,

$$\operatorname{tr}(\alpha'_\mu(\delta\alpha_\lambda - c_\lambda \alpha_\lambda)) = 0.$$

By the non-degeneracy of the trace function:

$$\delta\alpha_\lambda = c_\lambda \alpha_\lambda.$$

Thus left multiplication by δ is represented by the diagonal matrix $\operatorname{diag}(c_1, \ldots, c_d)$. Hence $\operatorname{tr}(\delta) = \sum_{\lambda=1}^d c_\lambda$ and $N(\delta) = \prod_{\lambda=1}^d c_\lambda$, and the result follows from the relation between the arithmetic and geometric means of positive numbers. □

Lemma 4.4. In the notation of the last lemma, suppose that $\left(\frac{1}{d}\right)\operatorname{tr}(\delta) \leq CN(\delta)^{\frac{1}{d}}$, for some constant C. Then $\left(\frac{1}{\operatorname{tr}(\delta)}\right)\delta \in K$ for some compact set in $P(k_R)$ (C, K independent of δ of course).

Proof. Using the $\prec$ notation, the constants present depending only on k_R, the given inequality says $\sum_1^d c_\lambda \prec (c_1 \cdots c_d)^{\frac{1}{d}}$, and since $0 < c_1 \leq c_2 \leq \cdots \leq c_d$, this gives:

$$c_d < \sum_1^d c_\lambda \prec (c_1 \cdots c_d)^{\frac{1}{d}} \leq c_1^{\frac{1}{d}} c_d^{\frac{d-1}{d}} = c_d \left(\frac{c_1}{c_d}\right)^{\frac{1}{d}}.$$

Divide by c_d, and so $1 \prec \left(\frac{c_1}{c_d}\right)^{\frac{1}{d}}$ or:

$$c_d \prec c_1. \tag{3}$$

Now $\operatorname{tr}(\zeta'\delta\zeta) = \sum c_\lambda u_\lambda^2$, $\operatorname{tr}(\zeta'\zeta) = \sum u_\lambda^2$ and so

$$c_1 \operatorname{tr}(\zeta'\zeta) = c_1 \sum u_\lambda^2 \leq \sum c_\lambda u_\lambda^2 \leq c_d \sum u_\lambda^2 = c_d \operatorname{tr}(\zeta'\zeta).$$

Picking out the first, third, and fifth terms and using (3):

$$c_1 \operatorname{tr}(\zeta'\zeta) \leq \operatorname{tr}(\zeta'\delta\zeta) \leq c_d \operatorname{tr}(\zeta'\zeta) \prec c_1 \operatorname{tr}(\zeta'\zeta).$$

That is to say:

$$c_1\operatorname{tr}(\zeta'\zeta) \asymp \operatorname{tr}(\zeta'\delta\zeta). \tag{4}$$

On the other hand $dc_1 \le \sum_1^d c_\lambda \le dc_d \prec c_1$ (by (3)) and so

$$c_1 \asymp \operatorname{tr}(\delta).$$

Combining with (4) it follows that $\frac{\operatorname{tr}\left(\zeta'\left(\frac{\delta}{\operatorname{tr}\delta}\right)\zeta\right)}{\operatorname{tr}(\zeta'\zeta)}$ is bounded above and below by positive constants, say C_1, C_2. But the set of all positive symmetric $a \in k_R$ such that $C_1 \le \frac{\operatorname{tr}(\zeta' a\zeta)}{\operatorname{tr}(\zeta'\zeta)} \le C_2$ is a compact set K in $P(k_R)$, and it has been shown that $\frac{\delta}{\operatorname{tr}\delta} \in K$, as required. □

Corollary 4.5. The bounded property obtained above for $\frac{\operatorname{tr}\left(\zeta'\left(\frac{\delta}{\operatorname{tr}\delta}\right)\zeta\right)}{\operatorname{tr}(\zeta'\zeta)}$ can be stated in the form:

$$\operatorname{tr}(\zeta'\delta\zeta) \asymp \operatorname{tr}(\delta)\operatorname{tr}(\zeta'\zeta).$$

5 Proof of Theorem 4.1

The matrix b whose existence is asserted in the theorem is to be constructed by changing from the basis $\underline{e}_1,\ldots,\underline{e}_n$ (cf. §2) in k_R^n to a new basis $\underline{b}_1,\ldots,\underline{b}_n$ consisting entirely of elements of ϑ^n. Thus the two bases will be connected by equations $\underline{b}_i = \sum_{j=1}^n \underline{e}_j\beta_{ji}$, where $\beta_{ji} \in \vartheta$. The norm $N(b)$ of the matrix of transformation $b = (\beta_{ij})$ is computed by expressing the linear mapping $\underline{x} \to b\underline{x}$ as a linear transformation of k_R^n into itself, regarded as a vector space over R, and calculating the determinant of the resulting matrix. Remembering that a basis was chosen for k so that the structure constants would be integers, and that in terms of this basis the elements of ϑ are exactly those elements of k with integral components, it follows at once that the real matrix representing the mapping $\underline{x} \to b\underline{x}$ has rational integral elements. $N(b)$ is thus a rational integer. Also, since b is the matrix corresponding to a change of basis in k_R^n, $N(b) \neq 0$, and so $|N(b)| \ge 1$, an inequality which will be needed later.

The construction of the new basis $\underline{b}_1,\ldots,\underline{b}_n$ is carried out as follows. Proceeding inductively, suppose $\underline{b}_1,\ldots,\underline{b}_{i-1}$ have already been chosen as elements of ϑ^n which are linearly independent in k^n regarded as a right module over k, and let $V_{i-1} = \underline{b}_1 k + \cdots + \underline{b}_{i-1} k$. (In case $i = 1$, V_{i-1} is to be $\{0\}$.) Now define:

$$\mu_i = \left[\min_{\underline{x}\in\vartheta^n,\underline{x}\notin V_{i-1}}\right] A(\underline{x},\underline{x})$$

where, as usual $A(\underline{x},\underline{x}) = \operatorname{tr}(\sum \xi_i'\alpha_{ij}\xi_j)$, $(\alpha_{ij}) = a$ being a positive symmetric element of A_R, as in the statement of the theorem. Also $\underline{x} = \sum_{i=1}^n e_i\xi_i$, in the notation of §2. Let $\underline{b}_i$ be an element of ϑ^n such that $A(\underline{b}_i,\underline{b}_i) = \mu_i$.

It must be checked that $\underline{b}_1,\ldots,\underline{b}_i$ are linearly independent over k. If a linear relation over k existed between these elements, it could be divided by the coefficient of $\underline{b}_i$ (since k is a division algebra) and so it would follow that $\underline{b}_i \in V_{i-1}$, contrary to the definition of $\underline{b}_i$. And so, proceeding step by step till $i = n$, a basis of k^n over k is obtained. By the definition of ground field extension, the resulting $\underline{b}_i$ also form a basis of k_R^n over k_R.

Note incidentally that the μ_i, being minima of $A(\underline{x},\underline{x})$ under increasingly strict conditions, satisfy

$$0 \le \mu_1 \le \mu_2 \le \cdots \le \mu_n.$$

The main burden of the proof now consists in verifying that, b being the matrix representing the above change of basis, $N(b)$ is bounded and $b'ab$ is contained in a Siegel domain. The last requirement can be stated differently. For, writing $\underline{x} = \sum \underline{e}_i \xi_i = \sum \underline{b}_i \eta_i$, the η_i and ξ_i are connected by the equations $\xi_i = \sum_{j=1}^n \beta_{ij}\eta_j$, where $b = (\beta_{ij})$, and so $A(\underline{x},\underline{x})$, when expressed in terms of the η_i, has matrix $b'ab$. At the same time the Babylonian reduction can be applied to $A(\underline{x},\underline{x})$, starting from the expression in terms of the η_i. If the variables $\zeta_1,\ldots,\zeta_n$ are those introduced by the reduction process, then $A(\underline{x},\underline{x})$ will have the form $\sum_{i=1}^n \operatorname{tr}(\zeta_i'\delta_i\zeta_i)$ and the η_i and ζ_i will be connected by equations of the form:

$$\zeta_i = \eta_i + \tau_{i,i+1}\eta_{i+1} + \cdots + \tau_{in}\eta_n \qquad (i = 1,\ldots,n).$$

Thus the matrix t of coordinate transformation is triangular unipotent. And writing $d = \operatorname{diag}\{\delta_1,\ldots,\delta_n\}$ it is clear that $b'ab = t'dt$. Hence the condition that $b'ab$ is in a Siegel domain will be established by proving the appropriate inequalities on the τ_{ij} and δ_i.

The proofs of these inequalities are contained in the following sequence of lemmas. In carrying out the proofs the following auxiliary quadratic form will be used:

$$A_1(\underline{x},\underline{x}) = \sum_{i=1}^n \frac{1}{\mu_i}\operatorname{tr}(\zeta_i'\delta_i\zeta_i).$$

Lemma 5.1. *C being a constant depending only on the initial data*

$$N(b)^2 \cdot \prod_{i=1}^n \left(\frac{\mu_i^d}{N(\delta_i)}\right) \le C.$$

Proof. Lemma 4.2 is to be applied to the auxiliary form $A_1(\underline{x},\underline{x})$ introduced above. This is a positive definite quadratic form on the vector space having the basis $\{e_i\omega_\lambda\}$, $i = 1,\ldots,n$, $\lambda = 1,\ldots,d$, the e_i forming a basis of k_R^n over k_R, and the ω_λ a basis of k_R over R. In terms of this basis the elements of ϑ^n are exactly those which have rational integral coordinates. Thus k_R^n, with this basis, is to be identified with Euclidean space R^{nd}, and ϑ^n with the lattice L of points with integral coordinates. Applying Lemma 4.2, with $m = nd$,

$$\min A_1(\underline{x},\underline{x}) \le CD^{\frac{1}{nd}}, \tag{5}$$

the minimum being taken over non-zero $\underline{x} \in \vartheta^n$, C being a constant, and D the determinant of $A_1(\underline{x},\underline{x})$ when expressed as a polynomial in the coefficients of the basis elements $e_1\omega_\lambda$. On the other hand $A_1(\underline{x},\underline{x})$ is actually defined in terms of $\underline{z} = (\zeta_1,\ldots,\zeta_n)$, where $\underline{y} = t^{-1}\underline{z}$, and $\underline{x} = b\underline{y}$. Thus $A_1(\underline{x},\underline{x}) = \underline{x}'a_1\underline{x}$, say $= \underline{z}'(t^{-1})'b'a_1bt^{-1}\underline{z}$.

Remembering that $N(b)$ is the determinant of the transformation $\underline{x} = b\underline{y}$, expressed in terms of a real $nd \times nd$ matrix, and that the similar determinant corresponding to $\underline{y} = t^{-1}\underline{z}$ is 1 (since t is triangular unipotent) it follows that the determinant of $A_1(\underline{x},\underline{x})$, when expressed in terms of the coordinates of $\underline{z}$, is $DN(b)^2$.

Here the fact is used that linear substitution on a quadratic form multiplies the determinant of the form by the square of the determinant of the substitution. Now, in terms of $\underline{z}$, $A_1(\underline{x},\underline{x}) = \sum_{i=1}^{n} \frac{1}{\mu_i}\operatorname{tr}(\zeta_i'\delta_i\zeta_i)$. To compute the corresponding determinant, let $\zeta_i = \sum_{\lambda=1}^{d} \zeta_{i\lambda}\omega_\lambda$ and let $\delta_i\omega_\mu = \sum_{\nu=1}^{d} c_{i\mu\nu}\omega_\nu$ where $\zeta_{i\lambda}, c_{i\mu} \in R$. Note that by the definition of norm, $\det(c_{i\mu\nu}) = N(\delta_i)$ (where μ,ν are the row and column indices). Then $\zeta_i'\delta_i\zeta_i = \sum_{\lambda,\mu=1}^{d} \zeta_{i\lambda}\zeta_{i\mu}\omega_\lambda'\delta_i\omega_\mu = \sum_{\lambda,\mu,\nu=1}^{d} \zeta_{i\lambda}\zeta_{i\mu}c_{i\mu}\cdots\omega_\lambda'\omega_\nu$. And so the determinant of $\sum_{i=1}^{n} \frac{1}{\mu_i}\operatorname{tr}(\zeta_i'\delta_i\zeta_i)$, regarded as a quadratic polynomial in the $\zeta_{i\lambda}$, is:

$$\prod_{i=1}^{n} \frac{1}{\mu_i^d} \det(\operatorname{tr}(c_{i\mu\nu}\omega_\lambda'\omega_\nu)) = \prod_{i=1}^{n} \frac{1}{\mu_i^d} \det(c_{i\mu\nu})\det(\omega_\lambda'\omega_\nu)$$

(since the $c_{i\mu\nu} \in R$ can be taken outside the trace function; also the product rule for determinants is used). Replace $\det(c_{i\mu\nu})$ by $N(\delta_i)$ and note that $\det(\omega_\lambda'\omega_\nu)$ is an absolute constant in the usual sense, and the last expression becomes $C\prod_1^d \frac{1}{\mu_i^d} N(\delta_i)$, which is, as pointed out above, equal to $DN(b)^2$. Combining this with (5):

$$\min_{\underline{x}\in\vartheta^n, x\neq 0} A_1(\underline{x},\underline{x}) \le C'N(b)^{-\frac{2}{nd}} \prod_1^d \mu_i^{-\frac{1}{n}} N(\delta_i)^{\frac{1}{nd}}. \tag{6}$$

Now in (6), let $\underline{x}$ be a given value so that the minimum on the left is actually attained. Suppose that, in terms of the basis $\underline{b}_1,\ldots,\underline{b}_n$ of k_R^n, $\underline{x}$ has coordinates $(\eta_1,\ldots,\eta_i,0,\ldots,0)$, with $\eta_i \neq 0$. Thus $\underline{x} \in V_{i-1}$, and so $A(\underline{x},\underline{x}) \ge \mu_i$. Also, since the ζ_i are obtained from the η_j by means of a triangular matrix, $\zeta_j = 0$ for $j \ge i+1$. And so $A_1(\underline{x},\underline{x}) = \sum_{j=1}^{i} \frac{1}{\mu_j}\operatorname{tr}(\zeta_j'\delta_j\zeta_j) \ge \frac{1}{\mu_i}\sum_{j=1}^{i}\operatorname{tr}(\zeta_j'\delta_j\zeta_j)$ (since the μ_j increase with j) and so $A_1(\underline{x},\underline{x}) \ge \frac{1}{\mu_i}A(\underline{x},\underline{x}) \ge 1$. Combining this with (6):

$$1 \le C'N(b)^{-\frac{2}{nd}} \prod_1^d \mu_i^{-\frac{1}{n}} N(\delta_i)^{\frac{1}{nd}}.$$

And this can be rearranged to give the inequality in the statement of the lemma. □

Lemma 5.2. For each i, $\mu_i \ge \operatorname{tr}\delta_i$.

Proof. $\mu_i = A(\underline{b}_i,\underline{b}_i)$, where, by definition, the η-coordinates of $\underline{b}_i$ are $\eta_j = 0$ $(j \neq i)$ and $\eta_i = 1$. Transforming to the ζ-coordinates of $\underline{b}_i$ by means of the triangular unipotent matrix t, it follows that $\underline{b}_i$ has the ζ-coordinates $(\tau_{1i},\tau_{2i},\ldots,\tau_{i-1,i},1,0,\ldots,0)$. Then, since $A(\underline{x},\underline{x}) = \sum_1^n \operatorname{tr}(\zeta_i'\delta_i\zeta_i)$, it follows that:

$$\mu_i = A(\underline{b}_i,\underline{b}_i) = \operatorname{tr}(\delta_i) + \sum_{j=1}^{i-1} \operatorname{tr}(\tau_{ji}'\delta_j\tau_{ji}) \ge \operatorname{tr}(\delta_i)$$

(because the δ_j, being obtained by Babylonian reduction from a positive matrix are positive in k_R), and this is the required result. □

Lemma 5.3. $|N(b)|$ is bounded by a constant depending only on the initial data; $\frac{\delta_i}{\operatorname{tr}(\delta_i)}$ lies in a compact set in $P(k_R)$ and $\operatorname{tr}(\delta_i) \prec \operatorname{tr}(\delta_{i+1})$.

Proof. Combine Lemmas 4.3 and 5.2 obtaining:

$$\mu_i \geq dN(\delta_i)^{\frac{1}{d}}.$$

It follows that each factor $\frac{\mu_i^d}{N(\delta_i)}$ in Lemma 5.1 is bounded below. It has already been noted that $|N(b)|$ is bounded below, and so Lemma 5.1 implies that each factor $|N(b)|$ and $\frac{\mu_i^d}{N(\delta_i)}$ is bounded above. This gives the boundedness of $|N(b)|$ as required. The boundedness of $\frac{\mu_i^d}{N(\delta_i)}$ along with Lemmas 4.3 and 5.2 implies that $\frac{\mathrm{tr}(\delta_i)^d}{N(\delta_i)}$ is bounded above and below for each i, and consequently, by Lemma 4.4, $\frac{\delta_i}{\mathrm{tr}(\delta_i)}$, lies in some compact set in $P(k_R)$, as required. Since $\frac{\mu_i^d}{N(\delta_i)}$ is bounded above, and $N(\delta_i)^{\frac{1}{d}} \leq \frac{1}{d}\mathrm{tr}(\delta_i)$, $\mu_i \prec \mathrm{tr}\delta_i$. Combining this with Lemma 5.2, $\mu_i \asymp \mathrm{tr}(\delta_i)$. Then since $\mu_i \leq \mu_{i+1}$ for each i it follows that $\mathrm{tr}(\delta_i) \prec \mathrm{tr}(\delta_{i+1})$, as required. □

Corollary 5.4. It appeared at the end of the above proof that $\mu_i \asymp \mathrm{tr}(\delta_i)$. Combining this with the corollary to Lemma 4.4 it follows that:

$$\mathrm{tr}(\zeta'\delta_i\zeta) \asymp \mu_i \mathrm{tr}(\zeta'\zeta)$$

for each i and any $\zeta \leq k_R$.

The last lemma obtained most of the conditions necessary to make $b'ab$ lie in a Siegel domain. The remaining condition, on t will now be proved.

Lemma 5.5. The matrix t, in the notation of §4, lies in a compact set of T.

Proof. By the definition of $\underline{b}_i$,

$$A(\underline{b}_i, \underline{b}_i) \leq A(\underline{b}_i + \sum_{j=1}^{h} \underline{b}_j \eta_j, b_i + \sum_{j=1}^{h} \underline{b}_j \eta_j) \tag{7}$$

for any $\eta_1, \ldots, \eta_h$ with $h < i$. Now, transforming by t, the ζ-coordinates of $\underline{b}_i + \sum_1^h \underline{b}_j \eta_j$ are

$$\begin{aligned}
&\zeta_1 = \eta_1 + \tau_{12}\eta_2 + \cdots + \tau_{1h}\eta_h + \tau_{1i},\\
&\zeta_2 = \eta_2 + \tau_{23}\eta_3 + \cdots + \eta_{2h}\eta_h + \tau_{2i},\\
&\cdots\\
&\zeta_h = \eta_h + \tau_{hi},\\
&\zeta_j = \tau_{ji} (h < j < i),\\
&\zeta_i = 1,\\
&\zeta_j = 0, \quad j > i.
\end{aligned} \tag{8}$$

Then, expressing (7) in terms of the ζ_j, and noting that the left-hand-side is obtained from the right hand side by setting $\eta_1 = \cdots = \eta_h = 0$, it follows that

$$\sum_{j=1}^{n} \mathrm{tr}(\zeta_j' \delta_j \zeta_j) \geq \sum_{j=1}^{i-1} \mathrm{tr}(\tau_{ji}' \delta_j \tau_{ji}) + \mathrm{tr}(\delta_i).$$

Since $\zeta_j = \tau_{ji}$ for $h < j < i$ and $\zeta_i = 1$ and $\zeta_j = 0$ for $j > i$ some of the terms here cancel, leaving:

$$\sum_{j=1}^{h} \operatorname{tr}(\zeta'_j \delta_j \zeta_j) \geq \sum_{j=1}^{h} \operatorname{tr}(\tau'_{ji} \delta_j \tau_{ji}).$$

Since the terms on the right are all positive it follows that:

$$\operatorname{tr}(\tau'_{hi} \delta_h \tau_{hi}) \leq \sum_{j=1}^{h} \operatorname{tr}(\zeta'_j \delta_j \zeta_j).$$

Apply the corollary to Lemma 5.3 to both sides. Hence:

$$\mu_h \operatorname{tr}(\tau'_{hi} \tau_{hi}) \prec \sum_{j=1}^{h} \mu_j \operatorname{tr}(\zeta'_j \zeta_j)$$

$$\leq \mu_h \sum_{j=1}^{h} \operatorname{tr}(\zeta'_j \zeta_j).$$

And dividing by μ_h:

$$\operatorname{tr}(\tau'_{hi} \tau_{hi}) \prec \sum_{j=1}^{h} \operatorname{tr}(\zeta'_j \zeta_j). \tag{9}$$

Now in (8) the η_j are arbitrary. If they are taken in ϑ, it is clear that they can be chosen so that the coordinates of $\zeta_1, \dots, \zeta_h$ (in terms of the basis $\omega_1, \dots, \omega_d$) are bounded. Explicitly η_h is chosen so that ζ_h has the coordinates of τ_{hi}, reduced mod 1, then η_{h-1} so that ζ_{h-1} has the coordinates of $\tau_{h-1,h}\eta_h + \tau_{h-1,i}$ reduced mod 1, and so on. When this has been done it is clear that the right hand side of (9) will be bounded by an absolute constant, and so $\operatorname{tr}(\tau'_{hi}\tau_{hi})$ is bounded. This completes the proof of the lemma, and so that of Theorem 4.1.

A sketch of the method used in the proof of the classical case will now be given. Here $k = Q$, $k_R = R$, $\vartheta = z$. The procedure is essentially as described for the general case, with the following modification in the construction of the new basis $\underline{b}_1, \dots, \underline{b}_n$ for $k_R^n = R^n$. The idea is to choose the basis this time so that it will actually form a basis for $\vartheta^n = Z^n$, with immediate consequence that $N(b) = \pm 1$ in the case. The $\underline{b}_i$ are to be constructed inductively. Suppose $\underline{b}_1, \dots, \underline{b}_{i-1}$ have already been chosen and that they form part of a basis $\underline{b}_1, \dots, \underline{b}_{i-1}, b'_i, \dots, \underline{b}'_n$ of Z^n. Set $V_{i-1} = \underline{b}_1 R + \cdots + \underline{b}_{i-1} R$. Now let $\underline{b}_i$ be an element of R^n such that $A(\underline{b}_i, \underline{b}_i) = \mu_i$ is the minimum of $A(\underline{x}, \underline{x})$ where $\underline{x}$ is allowed to vary over all elements of Z^n not in V_{i-1} such that $\underline{b}_{11}, \dots, \underline{b}_{i-1}, \underline{x}$ is part of a basis of Z^n. An alternative way of saying this is, if $\underline{b}_1, \dots, \underline{b}_{i-1}, \underline{b}'_i, \dots, \underline{b}'_n$ is a basis of Z^n, then $x = u_1 \underline{b}_1 + \cdots + u_{i-1} \underline{b}_{i-1} + u_i \underline{b}'_i + \cdots + u_n \underline{b}_n$, where the u_i are all in Z and $u_i, \dots, u_n$ have no common factor. Then the $\underline{b}_i$ so constructed form a basis of Z^n, and so $N(b) = \pm 1$. The rest of the proof is essentially as before. □

6 Siegel's Theorem

The Siegel-Minkowski theorem (Theorem 4.1) shows that the integral matrices in A_R operate on $P(A_R)$, by the transformations $a \to b'ab$ where $b \in M_n(\vartheta)$, $a \in P(A_R)$, in such a way that every a can be carried into a Siegel domain $\mathscr{S}$ by a matrix b of bounded norm, the bound being independent of a. Attention is now to be fixed on the effect of transformation of the type $a \to b'ab$ on the elements of a Siegel domain $\mathscr{S}$. Specifically it is to be shown that, if a bound is preassigned for the norm of b, and if $\mathscr{S}$ is any Siegel domain, then there are only finitely many matrices $b \in M_n(\vartheta)$ which will transform some $a \in \mathscr{S}$ into some other $a \in \mathscr{S}$. This is Siegel's theorem.

The proof will be carried out in two main stages, first of all for the classical case, in which the algebra k is taken to be Q, and then for the general case. The classical case itself will be broken down into a number of lemmas. The classical case of the theorem will first be stated explicitly:

Theorem 6.1. In the space of positive definite symmetric $n \times n$ matrices with real elements, let $\mathscr{S}$ be a Siegel domain and let L be a preassigned positive number. Write ${}^tM\mathscr{S}M$ for the set of all matrices ${}^tMAM$, $A \in \mathscr{S}$. Then there are only finitely many $n \times n$ matrices M with integral elements and $1 \le |\det M| \le L$ such that ${}^tM\mathscr{S}M \cap \mathscr{S} \neq \emptyset$.

The idea of the proof is to show that any matrix M having the properties just stated must have bounded elements, from which it will follow that there are only finitely many such matrices (their elements being integers). The proof will be by induction on n. If M satisfies the conditions stated in the theorem two cases are possible. M may be partitionable in the form $\begin{pmatrix} M_1 & N \\ 0 & M_2 \end{pmatrix}$ where M_1 and M_2 are square and of order less than n, or this may be impossible. In the first case it will be shown that M_1, M_2 have properties similar to M for Siegel domains of matrices of lower order, and that the elements of N are bounded in terms of M_1 and M_2. It then follows from the induction hypothesis that all the elements of M are bounded. In the second case a bound for the elements of M is found by a comparison of the orders of magnitude of a form and its Babylonian reduction.

Throughout the next four lemmas, then, it will be assumed that $M = \begin{pmatrix} M_1 & N \\ 0 & M_2 \end{pmatrix}$ where M_1 is of order $p \times p$, say, M_2 of order $(n-p) \times (n-p)$ with $1 \le p \le n-1$. If ${}^tM\mathscr{S}M \cap \mathscr{S} \neq \emptyset$, there are matrices A, A' in $\mathscr{S}$ such that $A' = {}^tMAM$. Let A and A' be partitioned in the same way as M; $A = \begin{pmatrix} A_{11} & A_{12} \\ A_{21} & A_{22} \end{pmatrix}$, $A' = \begin{pmatrix} A'_{11} & A'_{12} \\ A'_{21} & A'_{22} \end{pmatrix}$ where A_{11}, A'_{11} are of order $p \times p$.

Lemma 6.2. A_{11} and A'_{11} belong to a Siegel domain $\mathscr{S}_1$ in the space of all positive definite symmetric matrices of order $p \times p$, and $A'_{11} = {}^tM_1A_{11}M_1$.

Proof. Since $A \in \mathscr{S}$, $A = {}^tTDT$ where T is unipotent triangular, and D is diagonal, say $\mathrm{diag}\{d_1,\dots,d_n\}$, and $d_i \prec d_{i+1}$ for each i, and the elements of T are bounded, the absolute constants involved being independent of A. Partition D and $T: D = \begin{pmatrix} D_1 & 0 \\ 0 & D_2 \end{pmatrix}$, $T = \begin{pmatrix} T_1 & U \\ 0 & T_2 \end{pmatrix}$. Here D_1 and D_2 are diagonal, T_1 and T_2 are unipotent triangular. The equation $A = {}^tTDT$ gives the result $A_{11} = {}^tT_1D_1T_1$. The bounds on the elements of T and the ratio d_i/d_{i+1} give corresponding bounds for T_1 and D_1, so defining a Siegel domain $\mathscr{S}_1$. And the equation just written implies $A_{11} \in \mathscr{S}_1$. A similar discussion of A'_{11}

gives $A'_{11} \in \mathscr{S}_1$. If the matrices in the equation $A' = {}^tMAM$ are written in partitioned form then it can be seen at once that $A'_{11} = {}^tM_1A_{11}M_1$. □

Lemma 6.3. Define $B_{22} = {}^tT_2D_2T_2$, $B'_{22} = {}^tT'_2D'_2T'_2$. Then B_{22}, B'_{22} belong to a Siegel domain $\mathscr{S}_2$ in the space of positive definite symmetric matrices of order $(n-p)\times(n-p)$ and $B'_{22} = {}^tM_2B_{22}M_2$.

Proof. Here T_2, D_2 are as above. Since $A' \in \mathscr{S}$, $A' = {}^tT'D'T'$, with the appropriate boundedness condition on T' and D'. The matrices T'_2 and D'_2 in this lemma are obtained by partitioning T', D' in the same way as T and D. Then the bounding conditions on T, D, T', D' restricted to T_2, D_2, T'_2, D'_2 define a Siegel domain $\mathscr{S}_2$, and B_{22}, B'_{22}, by their definitions, belong to it.

To show that $B'_{22} = {}^tM_2B_{22}M_2$, start with the equation ${}^tMAM = A'$. This can be written as ${}^tM{}^tTDTM = {}^tT'D'T'$ or

$$\sqrt{D'}^{-1}\cdot{}^tT'^{-1}\cdot{}^tM\cdot{}^tT\cdot\sqrt{D}\cdot\sqrt{D}\cdot TMT'^{-1}\cdot\sqrt{D'}^{-1} = I_n \tag{10}$$

where the square root of D (or D') is the matrix $\mathrm{diag}(\sqrt{d_1},\dots,\sqrt{d_n})$ (or $\mathrm{diag}(\sqrt{d'_1},\dots,\sqrt{d'_n})$); these square roots are real since the d_i (and d'_i) are all positive. Equation (10) says that $\sqrt{D'}TMT'^{-1}\sqrt{D'}^{-1}$ is orthogonal. All the matrices in this product have a block of zeros in rows $(p+1)$ to n and columns 1 to p, and so the same is true of the product. But since the last matrix is orthogonal it must be of the form $\begin{pmatrix} H_1 & 0\\ 0 & H_2\end{pmatrix}$ where H_1 and H_2 are orthogonal and of order $p\times p$ and $(n-p)\times(n-p)$ respectively. Thus

$$\begin{pmatrix}\sqrt{D_1} & 0\\ 0 & \sqrt{D_2}\end{pmatrix}\begin{pmatrix}T_1 & U\\ 0 & T_2\end{pmatrix}\begin{pmatrix}M_1 & N\\ 0 & M_2\end{pmatrix}\begin{pmatrix}T_1'^{-1} & X\\ 0 & T_2'^{-1}\end{pmatrix}\begin{pmatrix}\sqrt{D'}^{-1} & 0\\ 0 & \sqrt{D'_2}^{-1}\end{pmatrix} = \begin{pmatrix}H_1 & 0\\ 0 & H_2\end{pmatrix} \tag{11}$$

where the fourth factor on the right is the partitioned form of T'^{-1}. (11) implies at once that $\sqrt{D_2}T_2M_2T_2'^{-1}\sqrt{D'_2}^{-1}$ is orthogonal, and writing this condition out explicitly (as in (10) with the suffix 2 attached) it follows at once that $B'_{22} = {}^tM_2B_{22}M_2$, as required. □

Lemma 6.4. In the notation of the last two lemmas:

$$N = -T_1^{-1}UM_2 + M_1T_1'^{-1}U'. \tag{12}$$

Proof. Starting with equation (11) in the proof of Lemma 6.3, compute the matrix on the left contained in row 1 to p and columns $p+1$ to n, and write down the condition that it should be zero. This turns out to be:

$$\sqrt{D_1}\cdot T_1\cdot M_1\times\sqrt{D'_2}^{-1} + \sqrt{D_1}(T_1N+UM_2)T_2'^{-1}\sqrt{D'_2}^{-1} = 0.$$

This is to say:

$$T_1M_1X + T_1NT_2'^{-1} + UM_2T_2'^{-1} = 0.$$

Solve this equation for N:

$$N = -M_1 X T_2' - T_1^{-1} U M_2. \tag{13}$$

Writing out explicitly the condition that $\begin{pmatrix} T_1'^{-1} & X \\ 0 & T_2'^{-1} \end{pmatrix}$ is the inverse of $\begin{pmatrix} T_1' & U' \\ 0 & T_2' \end{pmatrix}$ it turns out that $T_1' X + U' T_2'^{-1} = 0$, or $X T_2' = -T_1'^{-1} U'$. Substituting this in (13), the expression (12) is obtained. □

Lemma 6.5. Assuming a partitioning of M as above, the induction hypothesis implies that the elements of M are bounded by constants depending only on $\mathscr{S}$ and the upper bound L preassigned for $|\det M|$.

Proof. $\det M = \det M_1 \cdot \det M_2$, thus L is an upper bound for $\det M_1$ and $\det M_2$ (all the elements of M are integers). Also the constants defining the Siegel domains $\mathscr{S}_1$ and $\mathscr{S}_2$ in Lemmas 6.2 and 6.3 depend only on $\mathscr{S}$. By these lemmas ${}^t M_1 \mathscr{S}_1 M_1 \cap \mathscr{S}_1$ and ${}^t M_2 \mathscr{S}_2 M_2 \cap \mathscr{S}_2$ are not empty, and so the elements of M_1 and M_2 are bounded in the manner required. Then all the terms on the right of (12) in Lemma 6.4 are bounded (note that the elements of T_1^{-1} and $T_1'^{-1}$ are polynomials in those of T_1 and T_1', since these are triangular unipotent), and so all the elements of M are bounded.

This completes the step of the induction from orders $< n$ to n in the case where M has the special partitioned form given above. To complete the proof it must be shown that the elements of M are bounded when no such partitioning exists. □

Lemma 6.6. Let $A = {}^t TDT$ be in the Siegel domain $\mathscr{S}$, where T is unipotent triangular and D is diagonal, with the usual boundedness conditions. Writing x for the matrix of the column with elements $x_1, \ldots, x_n$, the relation ${}^t xAx \asymp {}^t xDx$ holds for all sets x_i not all zero.

Proof. In matrix notation, write $z = \sqrt{D}Tx$, $y = \sqrt{D}x$. Then ${}^t xAx = {}^t zz$ and ${}^t xDx = {}^t yy$. Thus ${}^t yy$ and ${}^t zz$ are to be compared where $z = \sqrt{D}T\sqrt{D}^{-1} y = Uy$, say. Here all the diagonal element u_{ii} are 1 and the only other non-zero u_{ij} are such that $i < j$. In this case $u_{ij} = \sqrt{d_i}\, t_{ij} \sqrt{d_j}^{-1}$. The t_{ij} are bounded and $d_i \prec d_j$ (for $i < j$), and so the u_{ij} are bounded. It follows at once that ${}^t zz \prec {}^t yy$. On the other hand $y = U^{-1} z$, where $U^{-1} = \sqrt{D}T^{-1}\sqrt{D}^{-1}$, and the elements of T^{-1} (also a triangular unipotent matrix) are polynomials in the t_{ij} and so are bounded. The above argument can be applied to give ${}^t yy \prec {}^t zz$, and so ${}^t yy \asymp {}^t zz$, as was to be shown. □

Corollary 6.7. By setting all the x_i equal to zero except one, the above lemma gives $a_{ii} \asymp d_i$ for each i.

Lemma 6.8. Returning to the notation of Theorem 6.1, suppose now that M is not partitionable as in Lemmas 6.2 to 6.5. Then for $1 \le p \le n-1$, $d_p \prec d_{p+1} \prec d_p' \prec d_{p+1}'$.

Proof. The hypothesis on M means that for each p $(1 \le p \le n-1)$ there is an element $m_{i(p)h(p)}$ of M with $i(p) \ge p+1$, $h(p) \le p$ and such that $m_{i(p)h(p)} \ne 0$. Now $A' = {}^t T' D' T'$, and so, by the corollary of Lemma 6.6,

$$d_i' \asymp a_{ii}' \tag{14}$$

for each i. But $A' = {}^tMAM$ and so $a'_{ii} = \sum_{j,h=1}^n a_{jh}m_{ji}m_{hi} \asymp \sum_{j=1}^n d_j m_{ji}^2$ (by Lemma 6.6). And so, since all the terms in the last summation are positive, $a'_{ii} \succ d_j m_{ji}^2$. Combine with (14):

$$d'_i \succ d_j m_{ji}^2. \tag{15}$$

For each i, j such that $m_{ji} \neq 0$ this means $d'_i \succ d_j$. In particular $m_{i(p)h(p)} \neq 0$ and so (15) becomes

$$d'_{h(p)} \succ d_{i(p)}. \tag{16}$$

Remembering that $h(p) \leq p < i(p)$ and that the order of magnitude of the d_i and d'_i increases with i, it follows that

$$\begin{aligned} d'_p \succ d'_{h(p)} &\succ d_i(p) \qquad \text{by (16)} \\ &\succ d_{p+1} \succ d_p \end{aligned} \tag{17}$$

and this is the required result. □

Lemma 6.9. With the notation and conditions of Lemma 6.8, all the elements of M are bounded.

Proof. $A' = {}^tMAM$, and so $A = {}^tM^{-1}A^1M^{-1}$. Write $m = \det M$ and $M' = mM^{-1}$; this is a matrix with integral elements. Then $m^2A = {}^tM'A'M'$. Thus m^2A and A' are related in the same way as A' and A, and so Lemma 6.8 can be applied. Note here that m^2A is in the same Siegel domain $\mathscr{S}$ as A, and m^2 is bounded by an absolute constant. Thus, Lemma 6.8, $\alpha'_p \prec d_p$. Then combining this with the actual statement of the last lemma:

$$d'_p \asymp d_p \qquad (1 \leq p \leq n).$$

Combine this with Lemma 6.8:

$$d_p \succ d_{p+1} \qquad (1 \leq p \leq n-1).$$

But $d_{p+1} \succ d_p$ and so $d_p \asymp d_{p+1}$. This means that, for all i, j:

$$d_i \asymp d_j. \tag{18}$$

On the other hand, it appeared in the proof of Lemma 6.8 (see (15)) that $d'_i \succ d_j m_{ji}^2$, and so, by (18)

$$d_i \succ d_j m_{ji}^2. \tag{19}$$

But, since $d_i \asymp d_j$ for all i, j, (19) means that m_{ji} is bounded.

The proof of Lemma 6.9, and with it that of Theorem 6.1, is thus completed. □

Theorem 6.1 must now be generalized to the corresponding result on Siegel domain in $P(A_R)$ where A as any simple algebra involution over Q. The result is:

Theorem 6.10. In the notation of the beginning of this section, let $\mathscr{S}$ be a Siegel domain in $P(A_R)$ and let L be a preassigned positive number. Then there are at most finitely many elements b of $M_n(\vartheta)$ such that $0 < |N(b)| \leq L$ and $b'\mathscr{S}b \cap \mathscr{S} \neq \emptyset$.

Proof. This result will be proved by introducing a basis for k_R over R, so that all the algebra elements are replaced by matrices with real entries, and in particular elements of $M_n(\vartheta)$ by matrices with rational integral elements. It will then be checked that the Siegel domain $\mathscr{S}$ goes over into a Siegel domain in the classical sense, and the result will follow by applying Theorem 6.1.

To carry out the above idea, take (as in §4) the basis $\omega_1,\ldots,\omega_d$ of k over Q so that $\vartheta = Z\omega_1+\cdots+Z\omega_d$ and $k_R = R\omega_1+\cdots+R\omega_d$. Correspondingly there is a basis, the $\underline{e}_i\omega_\lambda$, of k_R^n over R. Now, with respect to this basis, the linear transformation $\underline{x}\to b\underline{x}$ ($\underline{x}\in k_R^n$) has a matrix $\Phi(b)$ of order $nd\times nd$ over R, and, because of the way ϑ is defined in terms of the ω_λ, $\Phi(b)\in M_{nd}(Z)$ when $b\in M_n(\vartheta)$.

At the same time, the element $a\in M_n(k_R)$ defines a quadratic form $\mathrm{tr}(\underline{x}'a\underline{x})$. In terms of the basis $\{\underline{e}_i\omega_\lambda\}$ of k_R^n, this form (regarded as a quadratic form on k_R^n over R) will have a matrix $\Psi(a)$, say, and it is clear that $\Psi(b'ab) = {}^t\Phi(b)\Psi(a)\Phi(b)$.

The following lemmas will establish that the functions Φ and Ψ reduce Theorem 6.10 to Theorem 6.1. □

Lemma 6.11. The set of $b\in M_n(\vartheta)$ is carried by Φ into $M_{nd}(Z)$ and if $0<|N(b)|\le L$, then $0<|\det\Phi(b)|\le L$.

Proof. It has already been noted that the first part of this lemma holds. And the definition of $N(b)$ is simply $\det\Phi(b)$. □

Lemma 6.12. If t is unipotent triangular, then $\Phi(t)$ is unipotent triangular, and if t is in a compact set in T, $\Phi(t)$ is in a compact set in the appropriate space.

Proof. This is clear from the fact that Φ simply replaced each element of a matrix in $M_n(k_R)$ by a $d\times d$ real matrix, and in particular, the element 1 by the d-rowed identify matrix. Also the boundedness of elements of t is defined by that of the coordinates in terms of some basis, and so is equivalent to the boundedness of the elements of $\Phi(t)$. □

Lemma 6.13. Let δ be a positive symmetric element of k_R, and denote by $\Psi(\delta)$ the matrix (symmetric and positive definite) of $\mathrm{tr}(\xi'\delta\xi)$ expressed as a quadratic form on k_R over R, in terms of the basis $\omega_1,\ldots,\omega_d$. Then if δ is in a compact set K of $P(k_R)$ not containing zero, the Babylonian reduction of $\Psi(\delta)$ is carried out by a unipotent triangular matrix with bounded elements, and the diagonal elements of the reduced matrix will be bounded, both sets of bounds depending only on K.

Proof. This follows from the fact that all the construction of the Babylonian decomposition depends continuously on the initial matrix. □

Proof of Theorem 6.10 (cont'd.). Let a be in the given Siegel domain. Then $a = t'dt$ with t triangular and unipotent in a compact set of T and $d = \mathrm{diag}\{\delta_1,\ldots,\delta_n\}$, where $\mathrm{tr}\delta_i \prec \mathrm{tr}\delta_{i+1}$ and $\frac{\delta_i}{\mathrm{tr}\delta_i}$ is in a compact set of $P(k_R)$, a set which does not contain zero (cf. Lemma 5.3). Set $\delta_i = (\mathrm{tr}\delta_i)\delta_i^0$. Then $\Psi(d) = \begin{pmatrix} (\mathrm{tr}\delta_1)\Psi(\delta_1^0) & & \\ & \ddots & \\ & & (\mathrm{tr}\delta_n)\Psi(\delta_n^0)\end{pmatrix}$ in the notation of Lemma 6.13. It follows at once from that lemma that $\Psi(d) = {}^tTDT$ where T is a triangular unipotent matrix with real bounded elements and D is a diagonal matrix $\mathrm{diag}\{D_1,\ldots,D_{nd}\}$ satisfying $D_i \prec D_{i+1}$ for each i. Since $\Psi(a) = {}^t\Phi(t)\Psi(d)\Phi(t)$, the result just proved, along with Lemma 6.12 implies that $\Phi(a)$ is in a Siegel domain $\mathscr{S}$ of real symmetric $nd \times nd$ matrices. By Theorem 6.1, only finitely many integral matrices M with $0 < |\det M| \le L$ are such that ${}^tM\widehat{\mathscr{S}}M \cap \widehat{\mathscr{S}} \ne \emptyset$ and so certainly only finitely many M of the form $\Phi(b)$ with $0 < |N(b)| \le L$ and $b \in M_n(\vartheta)$. Since the mapping $\Phi : M_n(\vartheta) \to M_{nd}(Z)$ is one-one into, it follows that only finitely many $b \in M_n(\vartheta)$ with $0 < |N(b)| \le L$ are such that $b'\mathscr{S}b \cap \mathscr{S} \ne \emptyset$, as required. □

The following corollary of Theorem 6.10 is a variation of this theorem which will be required in §10. Assume that $\alpha \to \alpha^*$ is a second involution on k, extended by the transposition of matrices to A, and finally A_R, acting continuously on A_R and commuting with $x \to x'$.

Corollary 6.14. If $a \in \mathscr{S}$, a given Siegel domain, and C is a given constant, then there are at most finitely many elements b of $M_n(\vartheta)$ such that $0 < |N(b)| \le C$ and $b'(a^*)^{-1}b = a$.

Proof. a is in $\mathscr{S}$ and so $a = t'td$, where t is unipotent triangular with elements in a compact set M_1 and $d = \mathrm{diag}\{\delta_1,\ldots,\delta_n\}$ where $\mathrm{tr}\delta_i \prec \mathrm{tr}\delta_{i+1}$ and $\frac{\delta_i}{\mathrm{tr}\delta_i}$ is in a compact set M_2 not containing zero, for each i. Assume that M_1 and M_2 are enlarged if necessary to be invariant under $*$. Then $a^* = t^*d^*t'^* = (t'^*)'d^*(t'^*)$. It follows at once that $a^* = a_1$ is also in $\mathscr{S}$. Writing $a_1 = t_1'd_1t_1$, it is clear that $a_1^{-1} = t_1^{-1}d_1^{-1}t_1'^{-1} = (t_1'^{-1})'d_1^{-1}(t_1'^{-1})$ satisfies conditions similar to those of belonging to a Siegel domain, except that the orders of magnitude of the traces of the elements of d_1^{-1} are now decreasing instead of increasing, and $t_1'^{-1}$ has zeros above the diagonal instead of below. The boundedness conditions can be assumed to be the same as for $\mathscr{S}$, if necessary by modifying M_1, M_2 and the absolute constants involved. To get the elements of d_1^{-1} in the other order, and the triangular matrices the right way round, transform by the matrix $p = \begin{pmatrix} 0 & 0 & \cdots & 0 & 1 \\ 0 & 0 & \cdots & 1 & 0 \\ \vdots & \vdots & & \vdots & \vdots \\ 1 & 0 & \cdots & 0 & 0 \end{pmatrix}$.
Here $p' = p = p^{-1}$. Then $pa_1^{-1}p = pt_1^{-1}ppd_1^{-1}ppt_1'^{-1}p = (pt_1'^{-1}p)'(pd_1^{-1}p)(pt_1'^{-1}p)$ is in $\mathscr{S}$. The equation $b'(a^*)^{-1}b = a$ now becomes:

$$a = b'(a^*)^{-1}b = (pb)'pa_1^{-1}p(pb).$$

Thus pb transforms one element of $\mathscr{S}$ into another, and so, by Theorem 6.10, can take only finitely many values as was to be shown. □

7 The classical groups over a field

The groups to be described here are actually the restricted classical groups. That is to say they satisfy the extra condition of having their centers reduced to the identity. But, since this is the only kind of classical group to be discussed, the word "restricted" will be omitted.

The definitions will be given first over the complex numbers simply be giving a list of the groups, and then the case of other ground fields will be described.

The classical groups over the complex field C are to be the following types of groups:

(1) The projective group $P(n, C)$ $(n \geq 2)$ on n-variables. This is the quotient of the general linear group $GL(n, C)$ by its center, which consists of the group of non-singular scalar multiples of I_n.

(2) The projective orthogonal group $PO(n, C)$ $(n \geq 3)$ on n variables. This is the connected component of the identity in the quotient of the orthogonal group $O(n, C)$ (i.e., the subgroup of $GL(n, C)$ leaving the form $\sum_1^n x_i^2$ invariant) by its center. Here the center consists of the two matrices I_n and $-I_n$, where I_n is the n-rowed identity matrix. In taking the quotient group, two orthogonal matrices H and K map on the same element of $PO(n, C)$ if and only if $H = \pm K$. If n is even H and K will thus belong to the same component of $C(n, C)$ and so the two components will project separately into the quotient group. $PO(n, C)$ will in this case be understood to consist only of the connected component of the identity. If n is odd, the quotient of $O(n, C)$ by its center is connected and no further adjustment is needed.

(3) The symplectic group, reduced modulo its center, $Sp(n, C)$ with $n \geq 1$. The elements of this group (before reducing modulo the center) are linear transformations leaving an exterior form $\sum x_i \wedge y_i$ invariant.

(4) An direct product of groups of types (1), (2), (3).

This is the complete list. Note that all these groups are semi-simple connected Lie groups, the first three types being, in fact, simple.

Now for any subfield k of C, a classical group G over k will mean an algebraic group defined over k, such that, when the ground field k is extended to C, G becomes one of the above types up to isomorphism. The isomorphism will, of course, have to be expressed by means of rational functions over C.

The main object of this section is to show that the above groups can all be obtained as the connected components of the identity in automorphism groups over C of algebra with involutions. Here, by an automorphism of an algebra with involution, is understood an automorphism of the algebra in the ordinary sense which, in addition, commutes with the involution. The phrase "over C" means that the automorphisms in question all leave the elements of C invariant.

Consider first the algebra $M_n(C)$ of $n \times n$ matrices with complex elements $(n \geq 2)$. $M_n(C)$ is a simple algebra with center C (strictly speaking, the set of scalar matrices), and so, as is known from the theory of associative algebras (Skolem-Noether theorem), every automorphism α of $M_n(C)$ over C is an inner automorphism, that is to say given by $\alpha(X) = H^{-1} X H$, where $H \in M_n(C)$ is non-singular.

$M_n(C)$ has an involution, namely transposition. And since the product of two anti-automorphism is an automorphism, it follows that every anti-automorphism α of $M_n(C)$ is of the type $\alpha(X) = H^{-1} {}^t X H$, where t denotes transposition and $H \in M_n(C)$.

The main result of this section will be established, the way being prepared by the following lemmas.

Lemma 7.1. The anti-automorphism α, given by $\alpha(X) = H^{-1\,t}XH$ is an involution if and only if ${}^tH = \pm H$.

Proof. Using the definition of α twice, $\alpha\alpha(X) = H^{-1\,t}HX({}^tH)^{-1}H$. If α is an involution, the right hand side will be X, or in other words $({}^tH)^{-1}H$ commutes with X for all $X \in M_n(C)$. Hence $({}^tH)^{-1}H$ is a scalar multiple of I_n. That is ${}^tH = zH$, where $z \in C$. Transposing this equation and eliminating ${}^tH$, it follows that $H = z^2H$, or, since H is non-singular, $z = \pm 1$ as required. The converse is obvious. □

Lemma 7.2. Let α be as in Lemma 7.1 and let β be an automorphism of $M_n(C)$ over C, that is an inner automorphism, given by $\beta(X) = U^{-1}XU$. Then β commutes with α if and only if ${}^tUHU = zH$, $z \in C$.

Proof. Computing $\alpha\beta(X)$ and $\beta\alpha(X)$ it follows that, if $\alpha\beta = \beta\alpha$, then for all $X \in M_n(C)$, $U^{-1}H^{-1\,t}XHU = H^{-1\,t}U^tX({}^tU)^{-1}H$. This is the same as saying that $HU(({}^tU)^{-1}H)^{-1}$ commutes with ${}^tX$ for all $X \in M_n(C)$, and so is scalar. That is to say ${}^tUHU = zH$, for $z \in C$, as required. □

Lemma 7.3. If $M_n(C)$ is furnished with an involution, then the connected component of the identity of the group of automorphisms over C of the resulting algebra with involution is isomorphic either to $PO(n, C)$ or $Sp(n, C)$.

Proof. Let the involution α be as in Lemma 7.1. By Lemma 7.2 the automorphisms computing with α can be represented by matrices U such that ${}^tUHU = zH$. Since two matrices differing by a scalar factor induce the same inner automorphism, U can be replaced here by $\sqrt{z}U$. Thus U is to satisfy ${}^tUHU = H$. Here ${}^tH = \pm H$, that is, H is either symmetric of skew-symmetric. In the first case U represents a linear substitution leaving a quadratic form invariant (namely that with matrix H). Since, as mentioned, scalar multiples of a matrix induce the same inner automorphism as the matrix itself, it follows at once that the group of automorphisms is in this case $PO(n, C)$. If, on the other hand, ${}^tH = -H$, then ${}^tUHU = H$ means that U represents a substitution leaving a skew bilinear form invariant, and so, reasoning as before, the automorphism group is $Sp(n, C)$.

The group $P(n, C)$ can be obtained as the group of automorphisms of $M_n(C)$ with no involution. The following procedure, however, gives the group as automorphism group of an algebra with an involution. □

Lemma 7.4. Let A be the direct sum of two copies of $M_n(C)$, so that an element of A is a pair of matrices (X, Y). Let α be the involution given by $\alpha(X, Y) = ({}^tY, {}^tX)$. Then the connected component of the identity of the group of automorphisms over C of A with involution α is isomorphic to $P(n, C)$.

Proof. An automorphism β of A must either be an automorphism of each copy of $M_n(C)$ separately, or must, in addition, permute them. Here the former case must hold, since the connected component of the identity is wanted. The automorphism is over C, and so by the Skolem-Noether Theorem must induce an inner automorphism on each copy of $M_n(C)$. Thus β is of the form: $\beta(X, Y) = (U^{-1}XU, V^{-1}YV)$. It is easy to calculate as before that the condition for commuting with α is that $V = z\,{}^tU^{-1}$ for some $z \in C$. As before the scalar factor z is irrelevant and so β is determined by U. That means that the group of automorphisms over C of A with involution α is isomorphic to the group of automorphisms over C of $M_n(C)$, namely $P(n,C)$. □

The results of the above lemmas are now to be collected together:

Theorem 7.5. Let A be a semi-simple algebra over the complex field C, furnished with an involution α. Then the connected component G of the identity of the group of automorphisms over C of A, with the involution α, is a classical group over C. Conversely any such group can be obtained in this way.

Proof. The semi-simple algebra A can be written as a direct sum of simple components, each a matrix algebra over C. The involution α may act on some of these components as an involution and may permute others in pairs (as in Lemma 7.4). Write $A = A_1 + \cdots + A_r + B_1 + \cdots + B_s$, where α acts as an involution on A_i, a simple component, and where each B_i is a direct sum of two isomorphic simple components permuted by the involution. Then the automorphism group over C of A must either preserve all the simple components or permute them among themselves. Restricting attention to the connected component G of the identity, the first possibility is the only one. Now G is the direct product of the connected components of the identity in the automorphism groups of the A_i, B_j, each furnished with the appropriate involution. The above lemmas imply at once that G is a classical group over C.

Conversely if G is a classical group, each of its simple direct factors is the automorphism group (or connected component at least) over C of an algebra with involution. Then G is the connected component of the identity of the automorphism group over C of the direct sum of these algebras furnished with the sum of the individual involutions. □

8 Real forms of the classical groups

A classical group over R is by definition an algebraic group G defined over R such that, on extension of the coefficient field to C, G decomposes into a finite product of the classical groups over C described in §7. Each of these is the connected component of the identity of the automorphism group of an algebra with involution *over* C. The idea is to show that G can be obtained as the connected component of the identity of the automorphism group of an algebra with involution *over* R. To prove this it is necessary to use the fact that for the complex classical groups, each automorphism of the group is induced by an automorphism of the corresponding algebra with involution over C; and this is true except when one of the simple component groups is $PO(8)$.

Thus every classical group G over R, which when extended to a group over C does not contain $PO(8)$ as a component, is the connected component of the identity of the automorphism group of an algebra with involution over R.

In this section the simple classical group over R will be constructed explicitly as automorphism groups.

If the semi-simple algebra A over R is expressed as a direct sum of simple components then the involution on A may either act as an involution on some of the individual components or may permute some of the components in pairs. On the other hand, if attention is to be confined to the connected component of the identity of the automorphism groups, the automorphisms considered must all leave the components invariant.

There are thus two basic types to consider; namely a simple algebra A with involution, and an algebra $A = A_1 \bigoplus A_2$ where A_1 and A_2 are isomorphic simple algebras permuted by the involution. The various possible cases will now be considered individually.

Case 1. $A = A_1 \bigoplus A_2$, the A_i being simple algebras over R. The involution α on A will be of the form $\alpha(X_1, X_2) = (\beta(X_2), \gamma(X_1))$ where $X_1 \in A_1$, $X_2 \in A_2$. Since α^2 is the identity it follows at once that $\beta\gamma = \gamma\beta =$ the identity. Thus β is an arbitrary anti-isomorphism of A_2 onto A_1 and $\gamma = \beta^{-1}$.

Now let δ be an automorphism of A, carrying A_1 into itself and A_2 into itself. In particular δ must carry the center of A_i into itself ($i = 1, 2$). But A_i is a simple algebra over R and hence is of the form $M_n(R)$, $M_n(C)$ or $M_n(K)$ ($K =$ the quaternion algebra). The centers are respectively, isomorphic to R, C and R. In the second case the fact that the automorphism δ over R must belong to the connected component of the identity of the automorphism group implies that δ actually leaves C elementwise invariant. This happens automatically in the other two cases since δ is automorphism over R. Thus in any case δ acts as the identity on the center of A_i, and so by the Skolem-Noether Theorem δ is an inner automorphism. Thus $\delta(X_1, X_2) = (U_1^{-1}X_1U_1, U_2^{-1}X_2U_2)$, $U_1 \in A_1$, $U_2 \in A_2$. Computing the condition that $\alpha\delta = \delta\alpha$ it follows at once that $U_2 = z\gamma(U_1^{-1})$, where z is in the center of A_1. It follows at once that the automorphism δ is uniquely defined by the inner automorphism on A_1 induced by the matrix U_1. And so the connected component of the identity of the automorphism group of A with involution α is isomorphic to the group of inner automorphisms of A_1, namely to $P(n, R)$, $P(n, C)$, $P(n, K)$, according as $A_1 = M_n(R)$, $M_n(C)$ or $M_n(K)$.

In each of the next three cases A is taken as a simple algebra over R with an involution. Thus $A = M_n(R)$, $M_n(C)$ or $M_n(K)$ for some n.

Case 2. $A = M_n(R)$. The correspondence $X \to {}^tX$ is an involution, and the product of two anti-automorphisms is an automorphisms over R, and so an inner automorphisms. Hence the most general anti-automorphism α is of the type $\alpha(X) = A^{-1}{}^tXA$. For an involution, the condition $\alpha^2 =$ identity implies ${}^tA = zA$, $z \in R$, and so, as in Lemma 7.1, $z = \pm 1$. Thus $\alpha(X) = A^{-1}{}^tXA$ with ${}^tA = \pm A$. Continuing as in §7, it turns out that any automorphism over R, commuting with α, must be of the form $\beta : \beta(X) = U^{-1}XU$, where ${}^tUAU = zA$, $z \in R$. The corresponding automorphism group is thus orthogonal symplectic, according as ${}^tA = A$ or $-A$. Note that here, in the orthogonal case, there are several distinct possibilities, according to the index of inertia of the quadratic form of matrix A.

Case 3. $A = M_n(C)$. The involution α must be of the form $\alpha(X) = A^{-1}{}^tXA$ or $\alpha(X) = A^{-1\,t^{-}}XA$, according as the elements of the center are left invariant or are replaced by complex conjugates. Since attention is fixed on the connected component of the identity of the automorphism group of A, the automorphisms considered have the center invariant and so are inner. If $\alpha(X) = A^{-1}{}^tXA$, the condition for an involution gives

${}^{t}A = \pm A$, and the condition for an automorphism $\beta : \beta(X) = U^{-1}XU$ to commute with α is ${}^{t}UAU = A$. As in the complex case (§7) this yields two types of group, $PO(n, C)$ and $Sp(n, C)$.

On the other hand, if $\alpha(X) = A^{-1}\,{}^{t}\bar{X}A$, the condition $\alpha^2 =$ identity gives ${}^{t}\bar{A} = zA$ with $z \in C$. Transpose, take conjugates and eliminate A, where $z\bar{z} = 1$. Replacing A by $\frac{1}{\sqrt{z}}A$ (which does not affect α), it follows at once that ${}^{t}\bar{A} = A$. Then condition that the automorphism β, $\beta(X) = U^{-1}XU$, should commute with α is ${}^{t}UAU = A$ (where, if necessary U has been multiplied by a scalar factor), and so the corresponding group is the unitary group, modulo its center (which consists of multiples of the unit matrix by n-th roots of 1). There are different types corresponding to the index of inertia of A.

Case 4. $A = M_n(K)$. A has the involution $X \to {}^{t}\bar{X}$ (where the bar means the quaternionic conjugate), and the most general involution (reasoning as before) is α where $\alpha = A^{-1}\,{}^{t}\bar{X}A$, the condition for an involution being ${}^{t}\bar{A} = \pm A$. If ${}^{t}\bar{A} = A_1$, the corresponding automorphism group is of orthogonal type, leaving a quadratic (quaternionic) form invariant. This type is subdivided according to the index of inertia. In the case ${}^{t}\bar{A} = -A$, the corresponding automorphism group is of symplectic type, consisting of linear transformations leaving invariant a quaternionic bilinear form with matrix A. There is only one type of group here, for the matrix A may be brought, by a change of basis, into the form $\mathrm{diag}(i, i, \ldots, i)$, where i is one of the quaternion basis elements.

9 Algebras with involutions

The object of this section is to prove a number of preliminary lemmas to be used in deriving the main result of the next section. The first two are existence theorems for involutions satisfying additional conditions.

Lemma 9.1. A semi-simple algebra over the real number field always has a positive involution.

This was proved as Lemma 2.1 and is simply restated here for convenience.

Lemma 9.2. An involution on the center of a semi-simple algebra can always be extended to an involution of the whole algebra.

This is a classical result.

The next four lemmas are concerned with some deeper properties of positive involutions of algebras over R.

Lemma 9.3. If A is any semi-simple algebra over R with a positive involution α over R, then α maps each simple component of A onto itself.

Proof. Suppose α maps the simple component A_1 of A onto the component A_2, where $A_1 \neq A_2$. Then, if $a \in A_1$, $\alpha(a) \in A_2$, and since the decomposition of A into simple components is direct, $a\alpha(a) = 0$. On the other hand, since α is a positive involution $\mathrm{tr}(a\alpha(a))$ should be positive for non-zero a. This gives a contradiction. □

Lemma 9.4. Let A be any semi-simple algebra over R and let α, β be two positive involutions over R on A. Then there is an inner automorphism γ on A such that $\beta = \gamma\alpha$.

Proof. By Lemma 9.3, both α and β operate on each simple component of A separately, and so it will be sufficient to consider the case where A itself is a simple algebra over R. Then A is of the form $M_n(R)$ or $M_n(C)$ or $M_n(K)$. In the first and third cases, the center of A is R, and so α and β coincide in the center of A, namely operating there as the identity. In the case $A = M_n(C)$, the center of A is C, and α, being an involution over R, must either act as the identity on C or must induce the mapping $z \to z$. On the other hand $\text{tr}(z) = 2n^2$ (real part of z). And so, since α is positive, that is to say in particular $\text{tr}(z\alpha(z)) = 2n^2$ (real part of $z\alpha(z)$) is positive for $z \neq 0$, it follows that $\alpha(z) = \overline{z}$. Similarly $\beta(z) = \overline{z}$. Hence in all three cases α, β coincide on the center of A. Hence $\beta\alpha^{-1}$ is an automorphism on A leaving the center invariant, and so is an inner automorphism γ, as required. □

Lemma 9.5. Let A be a semi-simple algebra over R and let $x \to x'$ be a positive involution on A. Then if $a \in A$ is a positive symmetric element (cf. §2), there exits an element $b \in A$, positive and symmetric, such that $a = b^2$.

Proof. With respect to the bilinear form $\text{tr}(x'y)$ on A, the linear mapping $x \to ax$ is self-adjoint, and an orthogonal change of basis (i.e., orthogonal with respect to the form $\text{tr}(x'y)$) in A will bring the representative matrix of the mapping $x \to ax$ into diagonal form $D(a) = \text{diag}(d_1, d_2, \ldots, d_n)$. The d_i are all positive because a is positive. Now let P be a polynomial with real coefficients such that $P(d_i) = \sqrt{d_i}$ for each d_i. $b = P(a)$ is thus an element of A and it is clear that the representative matrix (in terms of the basis being used in A) of the mapping $x \to bx$ is $\sqrt{D(a)}$. It follows easily that $b^2 = a$. Also it is clear that b is symmetric and positive.

This proves the lemma; but it should be noted in addition that b is independent of the choice of the polynomial P. The element b so defined will be written as $a^{\frac{1}{2}}$. □

Lemma 9.6. Let $x \to x'$ be a positive involution on the semi-simple algebra A over R. Then $\text{tr}(ax'by)$ is a symmetric bilinear form in x and y with positive definite associated quadratic form if and only if $a = za_1$, $b = z^{-1}b_1$, where z is in the center of A and a_1 and b_1 are symmetric and positive.

Proof. Suppose the stated condition holds. The factor z will cancel out of the function $\text{tr}(ax'by)$, and so it is only necessary to prove the symmetry and positivity of this form when a and b are symmetric and positive. Using the fact that the trace is invariant under an involution and that factors under the trace function can be cyclically permuted:

$$\text{tr}(ax'by) = \text{tr}(byax') = \text{tr}(xa'y'b') = \text{tr}(xay'b) = \text{tr}(ay'bx).$$

This proves symmetry. Next, by Lemma 9.5, there are elements c, d in A such that $a = c^2, b = d^2, c = c', d = d'$. Hence $\text{tr}(ax'bx) = \text{tr}(ccx'ddx) = \text{tr}(cx'ddxc) = \text{tr}[(dxc)'(dxc)]$, and this is certainly positive for non-zero x.

Now the converse will be proved. It has been shown (Lemma 9.3) that a positive involution on A must map each simple component into itself. Also the direct decomposition of A into simple components induces a corresponding decomposition of the center, and multiplication is carried out componentwise. It will therefore be sufficient to prove the lemma for the simple components, or in other words, it can be assumed now for the rest of the proof that A itself is simple. Next suppose the lemma has been proved for one particular positive involution $x \to x'$; it will now be shown that the corresponding result will follow for any other positive involution α. By Lemma 9.4, α is given by the formula $\alpha(X) = u^{-1}x'u$, for some u in A. Since α is positive, $\mathrm{tr}(\alpha(x)y) = \mathrm{tr}(u^{-1}x'uy)$ is symmetric, with a positive definite associated quadratic form. And so, by the result as already assumed for $x \to x'$, u is a central element times a positive symmetric element. The central factor will cancel and so can be ignored, and so it can be assumed that $\alpha(x) = u^{-1}x'u$, with u symmetric and positive (with respect to the involution $x \to x'$). Now assume that $\mathrm{tr}(a\alpha(x)by)$ is symmetric with positive definite quadratic form. Applying the theorem, assumed for $x \to x'$, to this form, which can be written as $\mathrm{tr}(au^{-1}x'uby)$, it follows that $au^{-1} = za_1$, $ub = z^{-1}b_1$, where z is in the center and a_1, b_1 are symmetric and positive with respect to $x \to x'$. Thus $a = za_1u$, and $b = z^{-1}u^{-1}b_1$. It will be checked that a_1u and $u^{-1}b_1$ are symmetric and positive with respect to α. The symmetric is trivial. For positivity it must be shown that $\mathrm{tr}(u^{-1}x'ua_1ux) > 0$, for $x \neq 0$. Use Lemma 9.5 to write $u = v^2$, with $v' = v$. Then $\mathrm{tr}(u^{-1}x'ua_1ux) = \mathrm{tr}(v^{-1}v^{-1}x'ua_1ux) = \mathrm{tr}(v^{-1}x'ua_1uxv^{-1}) = \mathrm{tr}((uxv^{-1})'a_1(uxv^{-1})) > 0$, since a_1 is positive. Similarly $u^{-1}b_1$ is positive and symmetric with respect to α.

From what has just been shown, it is only necessary to prove the lemma for one particular positive involution. Now A is simple over R and so is of the form $M_n(R)$ or $M_n(C)$ or $M_n(K)$. The involution selected will be $x \to x' = {}^{t}\bar{x}$, where t denotes transposition and the bar denotes the identity in the first case and complex or quaternion conjugate in the second and third cases. This involution is positive (cf. §2).

With the meanings just explained for the notation, assume now that $\mathrm{tr}(ax'by)$ is symmetric with positive definite quadratic form. Using the symmetry of this form, and the properties of the trace:

$$\mathrm{tr}(ax'by) = \mathrm{tr}(ay'bx) = \mathrm{tr}(x'b'ya') = \mathrm{tr}(a'x'b'y).$$

Since the trace is non-degenerate, it follows that

$$ax'b = a'x'b', \qquad \text{i.e.,} \qquad a'^{-1}ax' = x'b'b^{-1}$$

for all x'. For $x' = 1$, this gives $a'^{-1}a = b'b^{-1}$, and so the common value of these is in the center. That is to say $b' = zb$, $a' = z^{-1}a$, for z in the center of A. Clearly $zz' = 1$. When the center is R (i.e., $A = M_n(R)$ or $M_n(K)$) this means $z = \pm 1$. Otherwise, if the center is C, $z' = \bar{z}$, and $z\bar{z} = 1$. In this case if b and a are replaced by $\sqrt{z}b$ and $\frac{1}{\sqrt{z}}a$, it turns out that $a' = a$, $b' = b$. Thus in all cases a and b can be taken to be both symmetric or skew-symmetric. Now let u be a matrix of n rows and one column with elements in R, C or K, according to the form of A and let $u' = {}^{t}\bar{u}$, where the bar has the appropriate meaning. Now set $x = y = uu'$. $\mathrm{tr}(ax'bx)$ must be positive, and it has in this case the value $\mathrm{tr}[(u'au)(u'bu)]$. If $A = M_n(R)$, $\mathrm{tr}[(u'au)(u'bu)] = (u'au)(u'bu)$, and for $x \neq 0$, that is $u \neq 0$, this could be zero if a and b were skew symmetric. So they must be symmetric, and both either positive or negative definite matrices. If $A = M_n(C)$, the trace of an element is twice the real part of the trace in the matrix sense. Also, it will be noticed in the above discussion a and b are both symmetric in the complex case.

Then, setting $x = y = uu'$ as above, $\mathrm{tr}(ax'bx) = \mathrm{tr}[(u'au)(u'bu)] = 2(u'au)(u'bu)$, both factors being real. The product has to be positive and so a and b are both the matrices of positive or negative definite Hermitian forms as required.

Finally, if $A = M_n(K)$, replace x in $\mathrm{tr}(ax'bx)$ by uv', where u and v are single column matrices of quaternions. Then $\mathrm{tr}(ax'bx) = \mathrm{tr}(v'avu'bu)$. Fixing u, v write $v'av = \sigma$ and $u'bu = \tau$. Then if a, b are both skew symmetric, σ and τ are pure imaginary quaternions. The form $\mathrm{tr}(ax'bx)$ can be divided out by a positive number so that $N(\sigma)$ and $N(\tau)$ both become 1. Then replace u by λu where λ is a quaternion of norm 1. Thus $\overline{\lambda} = \lambda^{-1}$. The requirement that $\mathrm{tr}(ax'bx)$ should be positive reduces to $\mathrm{tr}(\sigma\overline{\lambda}\tau\lambda) > 0$ for all λ. But since σ, τ are pure imaginary, λ can be chosen so that $\overline{\lambda}\tau\lambda = \lambda^{-1}\tau\lambda = -\overline{\sigma}$, and $\mathrm{tr}(\sigma\overline{\lambda}\tau\lambda)$ would become negative. Hence a and b must both be symmetric, and clearly must be either positive or negative definite. This completes the proof. □

Corollary 9.7. It appeared in the proof of the above lemma that the inner automorphism relating two positive involutions (as in Lemma 9.4) can be assumed to be induced by an element positive symmetric with respect to either of them.

It is convenient at this stage to introduce a special type of norm on A with values in the center. Let $a = \sum_1^m a_m$ be the expression of $a \in A$ into components, $a_i \in A_i$, corresponding to the direct decomposition of $A = \sum_1^m A_i$ into simple components. In particular, for z in the center of A let $z = \sum_1^m z_i$, z_i in the center of A_i. Let N_i denote the reduced norm on A_i. Then $N_i(z_i) = z_i^{r_i}$ whenever z_i is real (the center of A_i is either R or C) for some integer r_i. Now define $\nu(a) = \sum_1^m |N_i(a_i)|^{\frac{1}{r_i}}$, this sum being regarded as an element of the center of A, with $|N_i(a_i)|^{\frac{1}{r_i}}$ as its i-th component. Then clearly $\nu(ab) = \nu(a)\nu(b)$ and if z is an element of the center of A which has real positive components, $\nu(z) = z$. A further property of ν is that it is invariant under all auto- and anti-auto morphisms of A. For each such mappings either leaves unchanged or at most permutes the terms in the sum $\sum_1^m |N_i(a_i)|^{\frac{1}{r_i}}$ (note that the r_i corresponding to isomorphic simple components of A are certainly equal).

Now, returning to Lemma 9.4, and the corollary to the last lemma, let α and β be positive involutions on the semi-simple algebra A over R. Then $\beta(x) = u^{-1}\alpha(x)u$ where u can be assumed to be positive and symmetric with respect to α. If also $\beta(x) = v^{-1}\alpha(x)v$, with v positive and symmetric, then $v = zu$, with z in the center of A. Since u and v are positive and symmetric it follows easily that, in the direct decomposition of A, z will have real positive components. Hence if $\nu(u) = \nu(v) = 1$ it follows that $\nu(z) = z$ must be 1. It follows that the condition $\nu(u) = 1$ determines u uniquely. This gives the following stronger form of Lemma 9.4.

Lemma 9.8. Let α, β be two positive involutions on the semi-simple algebra A over R. Then there is a unique u in A, positive and symmetric with respect to α, and satisfying $\nu(u) = 1$, such that $\beta(x) = u^{-1}\alpha(x)u$ for all $x \in A$.

Lemma 9.9. Let A be a semi-simple algebra with involution ρ over R. Then there exists a positive involution of A over R commuting with ρ.

Proof. Let α be any positive involution on A over R (Lemma 9.1). Then $\rho\alpha\rho$ is clearly a positive involution, and so there is a unique $u \in A$ which is positive and symmetric with respect to α, satisfies $\nu(u) = 1$, and is such that $\rho\alpha\rho(x) = u^{-1}\alpha(x)u$ (Lemma

9.8). Let $v = u^{\frac{1}{2}}$, then v is symmetric and positive with respect to α (Lemma 9.5), and so β, defined by $\beta(x) = v^{-1}\alpha(x)v$ is a positive involution. It will be shown to commute with ρ. In

$$\rho\alpha\rho(x) = u^{-1}\alpha(x)u \tag{20}$$

replace x by $\rho(x)$ and apply ρ to both sides. The result can be arranged as

$$\rho\alpha\rho(x) = \rho(u)^{-1}\alpha(x)\rho(u). \tag{21}$$

Replace x by u in (20): then $\rho\alpha\rho(u) = u^{-1}\alpha(u)u = u$, and so $\alpha\rho(u) = \rho(u)$. That is to say $\rho(u)$ is symmetric under α. To show that $\rho(u)$ is positive with respect to α, note that

$$\begin{aligned} \mathrm{tr}(\alpha(x)\rho(u)x) &= \mathrm{tr}(\alpha\rho(y)\rho(u)\rho(y)) \qquad \text{(replacing } x \text{ by } \rho(y)) \\ &= \mathrm{tr}(yu\rho\alpha\rho(y)) \\ &= \mathrm{tr}(y\alpha(y)u) \qquad \text{(by (20))} \\ &= \mathrm{tr}(\alpha(y)uy) \end{aligned}$$

and this is positive, since u is positive with respect to α. Finally $\nu(\rho(u)) = \nu(u) = 1$. Thus u and $\rho(u)$ are both symmetric and positive with respect to α, and have $\nu = 1$, and so, by (20) and (21) and the uniqueness part of Lemma 9.8, $\rho(u) = u$. It follows, by the method of definition of $v = u^{\frac{1}{2}}$, that $\rho(v) = v$. Now, to show that ρ commutes with β, it must be proved that:

$$\rho(v)\rho\alpha(x)\rho(v)^{-1} = v^{-1}\alpha\rho(x)v. \tag{22}$$

The right hand side is

$$\begin{aligned} v^{-1}\alpha\rho(x)v &= v^{-1}\rho\rho\alpha\rho(x)v \\ &= v^{-1}\rho(u)\rho\alpha(x)\rho(u)^{-1}v \qquad \text{by (20)} \\ &= v^{-1}u\rho\alpha(x)u^{-1}v \qquad (\text{since } \rho(u) = u) \\ &= v\rho\alpha(x)v^{-1} \qquad (\text{since } u = v^2). \end{aligned}$$

And this is equal to the left hand side of (22) ($\rho(v) = v$) as required. □

Return now to the case of a simple algebra $A = M_n(k)$ over Q, furnished with an involution σ over Q. It is desirable for future working to express σ in a way which is related to the transposition operation on the matrices of $M_n(k)$. This is done in the next lemma.

Lemma 9.10. There exists an involution $\xi \to \xi^*$ on k, extended to A by setting $a^* = (\alpha_{ij})^* = (\alpha^*_{ji})$ such that $\sigma(a) = u^{-1}a^*u$ for some $u \in A$ such that $u^* = \pm u$.

Proof. By Lemma 9.2 there is an involution $*$ on k coinciding with σ on the center of k. Extend $*$ to A as stated in the lemma. Then, since the center of A consists of matrices of the type zI_n, with z in the center of k, it follows that σ and $*$ coincide on the center of A. Thus the mapping $x \to \sigma^{-1}(x^*)$ is an automorphism on A (over Q) inducing the identity on the center of A. This is therefore an inner automorphism and so $\sigma(x) = u^{-1}x^*u$ for some $u \in A$. Now $\sigma\sigma(x) = x$, and so it follows that $u^{-1}u^*$

commutes with x for all x. That is, $u^* = zu$, where z is in the center of A. Apply $*$ to this equation and eliminate u, where $zz^* = 1$. If the center is R this gives $z = \pm 1$. If $A = M_n(C)$, z^* may be z or $\overline{z}$. Hence $z = \pm 1$ or a factor $\sqrt{z}$ may be included in u so that $u^* = u$.

Both the involutions σ and $*$ can be extended to involutions of A_R over R, simply by coefficient extension. It will be assumed that this is now done, without any change in the notation.

Turning attention now to positive involutions, Lemma 9.9 asserts the existence of a positive involution $\xi \to \xi'$ on k_R, over R, commuting with $*$. This involution can be extended to an involution of A_R over R by setting $x' = (\xi_{ij})' = (\xi'_{ji})$. The extended involution on A_R is then positive (cf. remarks following Lemma 2.1) and clearly commutes with $*$. The idea now is to express, in terms of $*$ and the involution $x \to x'$, the family of all positive involutions on A_R which commute with the given involution σ. □

In this context it is convenient to modify Lemma 9.8. If α is any positive involution on A_R, then $\alpha(x) = a^{-1}x'a$, where a is uniquely determined by the conditions of being symmetric and positive (with respect to $x \to x'$) and $\nu(a) = \nu(u)$ (instead of 1), here u is the element of A appearing in Lemma 9.10. With these notations:

Lemma 9.11. The necessary and sufficient condition that α should commute with σ, where $\alpha(x) = a^{-1}x'a$, $\sigma(x) = u^{-1}x^*u$, is

$$u'a^{*-1}u = a.$$

Proof. Writing down the condition $\alpha\sigma = \sigma\alpha$ at once that $u'a^{*-1}u = za$, for some z in the center of A_R. It is easy to see that the symmetry and positivity of a imply those of a^{*-1}, and so z is symmetric and positive. But, since $\nu(a) = \nu(u)$, $\nu(z) = 1$, and so $z = 1$, as required. The converse is a straightforward calculation. □

The next two lemmas give a description of the family of all positive involutions commuting with σ. Let G be the connected component of the group of automorphisms of A_R commuting with σ. The restriction of G to the connected component implies that G transforms the simple components of A into themselves (cf. §8) and acts as the identity on the center. G thus consists entirely of inner automorphisms. If s is such an automorphism given by $s(x) = bxb^{-1}$, then it is easy to see that commutation of s with σ means $b^*ub = zu$ for some z in the center.

Now for any fixed positive involution α_0 commuting with σ, let K be the subgroup of G consisting of automorphisms commuting with α_0.

The condition for commutation with α_0 is $\alpha_0(b)b = z$ for some z in the center of A_R. It follows easily that the transformation $x \to bxb^{-1}$ leaves $\mathrm{tr}(\alpha_0(x)x)$ invariant. This is a positive definite form on the vector space underlying A_R, and so all such inner automorphisms form a subgroup of the orthogonal group (orthogonal with respect to the form $\mathrm{tr}(\alpha_0(x)x)$) on this vector space. The additional condition on K to be the set of inner automorphisms commuting with σ implies that K is a closed subgroup of this orthogonal group, and so K is compact. Now write $H = G/K$, the space of right cosets of K in G.

[Note. K is maximal compact subgroup of G and so G/K is the Riemann symmetric space associated with G if G has no invariant compact subgroups.]

Lemma 9.12. The set of positive involutions of A_R commuting with σ can be set in one-one correspondence with H.

Proof. If s is in G, then it is easy to see that $\alpha = s^{-1}\alpha_0 s$ is a positive involution commuting with σ. Also $s^{-1}\alpha_0 s = t^{-1}\alpha_0 t$ if and only if $ts^{-1} \in K$, that is to say, t, s belong to the same element of G/K. This gives a one-one mappings of H into the set of positive involutions of A_R commuting with σ. To check that the mapping is surjective, let α be a positive involution of A_R commuting with σ. By Lemma 9.8 $\alpha(x) = a^{-1}\alpha_0(x)a$, with a positive and symmetric with respect to α_0. By Lemma 9.5 write $a = b^2 = \alpha_0(b)b$, let s be the automorphism $s(x) = bxb^{-1}$. It is easy to check that $\alpha = s^{-1}\alpha_0 s$, and so the mapping described above will be surjective if s is in G; that is to say if s commutes with σ.

To prove this note first notice that the condition for α to commute with σ is $a\sigma(a) \in$ center, that is to say (in the notation of Lemma 9.10 etc.) $a^* ua = zu$, with z in the center of A_R. But this has been seen to be the condition for σ to commute with s_1, defined by $s_1(x) = axa^{-1}$. Clearly $s_1 = s^2$. Now, as in the definition of $a^{\frac{1}{2}}$ (Lemma 9.5) take the regular representation of A_R and choose a basis such that the representative matrix of a is diagonal. Then $x \to s_1(x)$ is a linear transformation of A_R. Replacing the algebra elements by their regular representation matrices, and regarding these as forming a subset of a full matrix algebra, s_1 is represented (with respect to the natural matrix basis) by a diagonal matrix with positive elements and similarly s by the square root of that matrix. Arguing as in Lemma 9.5, s can be written as a polynomial in s_1. s_1 commutes with α, and hence so does s. □

It will be noticed from the correspondence established in the last lemma that if the element s operates on H from the right, then, in the set of positive involutions on A_R commuting with σ, the operation is represented by the mapping $\alpha \to s^{-1}\alpha s$. An alternative form for this statement can be obtained by combining Lemmas 9.11 and 9.12, as follows.

Lemma 9.13. Let $\mathscr{H}$ be the set of positive symmetric (with respect to the involution $x \to x'$) elements a of A_R such that $\nu(a) = \nu(u)$ and $u'a^{*-1}u = a$. Then $\mathscr{H}$ and H can be set in one-one correspondence. Also if s is the element of G given by $s(x) = bxb^{-1}$, the operation of s on the right of H is represented on $\mathscr{H}$ by the mapping $a \to b'ab$.

Proof. The first part follows immediately from Lemmas 9.11, 9.12. Then if $\alpha(x) = a^{-1}x'a$, $s(x) = bxb^{-1}$, the operation of s on H is represented (see above) by the mapping $\alpha \to s^{-1}\alpha s$, and a computation shows at once that this is represented in $\mathscr{H}$ by the mapping $a \to b'ab$. □

10 *M*-domains

The concept about to be introduced finds its principal applications in the questions of compacity and measure-finiteness of the quotient space of a group with respect to a discrete subgroup.

Let G be an algebraic metric group, defined by polynomial equations over Q, but the elements of the matrices being allowed to take real (not necessarily rational) values.

Let G_Q be the subgroup of matrices in G with rational elements, and G_Z a subgroup of unimodular matrices in G with integral elements. Let K be a compact subgroup of G and write $H = G/K$. Then a set W in H will be called an M-domain for G_Z if:

(1) $G_Z W = H$;

(2) For each $s \in G_Q$, there are at most finitely many elements $\gamma \in G_Z$ such that $sW \cap \gamma W \neq \emptyset$.

The main task of this section is to prove the existence of such domains for the groups defined in §8 by means of algebras with involutions. Continuing with the notation of the last section A will be a simple algebra $M_n(k)$ over Q, k being a division algebra over Q, and A will be given with an involution σ. G is to be the connected component of the identity of the group of automorphisms of A_R over R which commute with σ. Each element of G, as has been pointed out, is an inner automorphism of A_R. Also, in terms of a basis for A over Q, taken also as a basis for A_R over R, the elements of G, which are linear transformations on A_R, can be represented by matrices. The conditions for being an automorphism of A_R and for commuting with σ are expressed by algebraic equations over Q in the elements of these matrices. Thus G is an algebraic matric group of the type described at the beginning of this section. As in §4 it will assumed that an order ϑ has been fixed in k and that the basis for k has been chosen as a basis of ϑ. The elements of $M_n(\vartheta)$ will, as before, be called integral.

With the matrix representation of G just described, it is clear that an element of G which is an inner automorphism of A_R induced by an element of A (i.e., an automorphism $x \to \omega^{-1}x\omega$ with ω in A) will be represented by a matrix with elements in Q. Conversely, suppose an element s of G has a matrix with rational elements. Since this matrix satisfies the algebraic conditions for representing an automorphism, which depend only on structural constants of A_R, that is of A, it follows that s is an automorphism of A. Since s is inner on A_R, it leaves the center of A elementwise fixed and so by the Skolem-Noether theorem, is an inner automorphism on A. That is to say $s(x) = \omega^{-1}x\omega$ for some ω in A. Summing up:

Lemma 10.1. In the notation used at the beginning of this section, G_Q consists of those elements of G which are automorphisms of A_R induced by elements of A.

The group G_Z is to be the set of inner automorphism in G of the form $x \to \omega x\omega^{-1}$, where ω and ω^{-1} are both integral. It is clear that in terms of the basis chosen for A_R, the elements of G_Z are represented by matrices with integral elements.

It turns out, incidentally, that G_Z is of finite index in the group of all matrices of G (in the representation just described) which have integral elements and determinant 1, but the proof of this point seems to be rather hard. No use will be made of this statement here, but it would be necessary if the invariance properties (up to commensurability) described in the introduction were being discussed.

The compact subgroup K of G used in the definition of an M-domain is to be taken as at the end of §9. Namely take a positive involution α_0 of A_R over R commuting with σ and let K be the subgroup of elements of G which commute with α_0. Then, writing $H = G/K$, Lemma 9.13 gives a characterization of H which will be used in what follows. The main result of this section can now be formulated:

Theorem 10.2. Let $A = M_n(k)$ be a simple algebra over Q with an involution σ (which may be replaced by the identity) and let G be the connected component of the identity of the group of automorphisms of A_R commuting with σ. Then G_Q, G_Z, K and H being as described above, H contains an M-domain for G_Z.

The remainder of this section will be occupied by the various steps of the proof. But first one or two remarks must be made. In the first place, a separate discussion will be required at each stage for the situation where σ is replaced by the identity, although this will often amount to noting that some condition is trivially satisfied (e.g., the condition on an automorphism or involution to commute with σ). The theorem could be formulated in a unified manner, with σ always an involution, by allowing A to be semi-simple. G would then be a direct product of groups, each of which would be as in Theorem 10.2 as stated. In particular A would be a direct sum of simple components, and σ would act on some of these new trivially, and permute others pairwise, the latter case (cf. Case 1 in §8) is equivalent to a simple algebra with no involution.

A final point to note here concerns the generality of Theorem 10.2. A_R is a semi-simple algebra over R, and so every G in Theorem 10.2 is a classical group over the real numbers (namely a direct product of types given in §8). Conversely all the classical groups can be obtained as automorphism groups of algebras with involutions with the exception noted at the beginning of §8.

The first step in the proof of Theorem 10.2 is to establish the following lemma:

Lemma 10.3. There is a finite set $b_1, \dots, b_N$ of elements of A, such that every integral b in A satisfying $0 < |N(b)| \le C$, for some fixed C, can be written in the form $b = gb_i$ for some i, where g belongs to the multiplicative group of integral invertible elements of A (i.e., elements g such that g and g^{-1} are both integral).

Proof. The proof will be made to depend on the special case in which $A = Q$, $\vartheta = Z$. It will be shown in fact that the set of all $n \times n$ matrices over Z satisfying $\det B = \pm\Delta$, for some fixed positive Δ, form a finite number of cosets of the group of unimodular matrices over Z in the group of all non-singular matrices over Q. Now any integral matrix B can be written as $B = XDY$ where D is a diagonal integral matrix and X, Y are unimodular integral matrices. $\det D = \pm\Delta$, and so there are only finitely many choices for D. It will therefore be sufficient to prove the result for the set of matrices B derived from a fixed diagonal D. Now there is a finite set of unimodular integral matrices such that any unimodular Y is congruent to one of them, say Y_ν, modulo Δ. Pick one of these Y_ν, and consider a $Y \equiv Y(\mathrm{mod}\,\Delta)$. Then $YY_\nu^{-1} = Z$ is an integral unimodular matrix $\equiv I_n(\mathrm{mod}\,\Delta)$ for this choice of Y_ν, with $Y \equiv Y_\nu(\mathrm{mod}\,\Delta)$, $B = XDY = XDZY_\nu = (XDZD^{-1})DY_\nu$. Here $XDZD^{-1}$ is integral (since $Z \equiv I_n(\mathrm{mod}\,\Delta)$) and unimodular, and so B is in the coset of the unimodular group, multiplied on the right by DY_ν. The DY_ν form a finite set, and so the result is proved for the special case.

To prove the lemma in the general case, note that the integral elements of A are represented in the regular representation (i.e., the matrix representation of the linear transformations $a \to ax$ corresponding to the appropriate choice of basis in A) by matrices with rational integral elements, and bounded norm means bounded determinants for the representative matrices; that is to say only a finite number of possible values for the determinant. Applying the above proof for each of these possible values $\pm\Delta$, the required result follows. □

The proof of Theorem 10.2 will be carried out by explicit construction, followed by verification of the condition to be an M-domain.

Continuing now with the proof of Theorem 10.2, let σ be an involution. In this case H will be represented as described in Lemma 9.13, namely as the set of all elements a of A_R positive and symmetric with respect to the involution $x \to x'$ and satisfying $\nu(a) = \nu(u)$ and $U'a^{*-1}u = a$. On the other hand if σ is the identity, the discussion of §9 can be carried through simply by omitting all mention of σ, and H will be represented by the set of elements a which are positive and symmetric (with respect to $x \to x'$) and satisfy $\nu(a) = 1$. In both cases, if an element s of G is an inner automorphism induced by b in A_R, the operation of s on H is represented by the correspondence $a \to b'ab$. Here b is arbitrary regular element of A_R if σ is the identity, but must satisfy a condition corresponding to the commutation of s and σ when the latter is non-trivial.

Now according to Theorem 4.1, there is a Siegel domain $\mathscr{S}$ and a constant C such that every a in $P(A_R)$ can be reduced to an element $b'ab$ in $\mathscr{S}$ by an integral element b of A_R such that $0 < |N(b)| \leq C$.

Now applying Lemma 10.3 to the set of $b's$ which reduce the elements of $P(A_R)$ to elements of $\mathscr{S}$, it follows that, for every $a \in P(A_R)$, there is a g which is integral and has integral inverse and an i such that $b_i'g'agb_i$ is in $\mathscr{S}$. Assuming for the moment that σ is the identity, define the set W by

$$W = \bigcup_i \{a : b_i'ab_i \in \mathscr{S}\}.$$

Thus the result just stated is that, for every $a \in P(A_R)$, there is an integral g with integral inverse such that $g'ag \in W$. In the case where σ is the identity, such an element g, operating in this way on a, corresponds to an element of G_Z operating on an arbitrary element of H. It has thus been shown that

$$G_Z W = H,$$

the first condition in the definition of an M-domain.

When σ is an involution the definition of W must be modified, as it is now not true that g necessarily represents the operation of an element of G_Z. g must now satisfy a condition corresponding to commutation with α. This is arranged in the following way. Take $a \in P(A_R)$, representing an element of H. a now satisfies $u'(a^*)^{-1}u = a$. Let b be an integral element of A such that $b'ab = a_0 \in \mathscr{S}$, $0 < |N(b)| \leq C$. Then, combining the equations $u'(a^*)^{-1}u = a$ and $b'ab = a_0$ it follows that

$$(b^*ub)'(a_0^*)^{-1}b^*ub = a_0.$$

Now apply the corollary of Theorem 6.10, noting that a bound for the norm of b implies a bound for that of b^*ub. It follows that there is a finite set $c_1, \dots, c_r$ in A_R such that, for every integral b satisfying $0 < |N(b)| \leq C$ and reducing some element of $P(A_R)$ to an element of $\mathscr{S}$, $b^*ub = c_i$ for some i.

Take each value of i in turn and consider, for that value of i, the set of all solutions of $b^*ub = c_i$. By Lemma 10.3, since the b are of bounded norm, there is a finite set of solutions b_j such that every solution can be written as gb_j where g and g^{-1} are integral. In particular $b_j^*g^*ugb_j = c_i = b_j^*ub_j$. Cancelling the b_j and b_j^* it follows that $g^*ug = u$.

Doing this for each of the c_i, a finite collection $b_1, \ldots, b_N$ of integral elements of A_R is obtained with the property that every element a of $P(A_R)$, satisfying $u'(a^*)^{-1}u = a$, can be reduced to an element of $\mathscr{S}$ by a transformation $a \to b_i' g' a g b_i$ for i, where g and g^{-1} are integral and $g^* u g = u$. It is easy to check that the last condition ensures that g represents an element of G_Z. Now define W as above, but using the new set of b_i, and it follows at once as before that $G_Z W = H$.

It will now be shown that, in each of the two cases considered W also satisfies the second condition for an M-domain. According to Lemma 10.1, the operation of an element of G_Q on H is represented by a transformation $a \to s' a s$ on the positive symmetric elements of A_R where s is in A. Now the operation of an element of G_Z is represented by a transformation $a \to g' a g$, where g and g^{-1} are integral elements of A. Thus it has to be shown that, for any $s \in A$ there are only finitely many integral elements g of A which have integral inverses such that a relation $s' a_1 s = g' a_2 g$ holds with a_1 and a_2 in W. This will now be proved. To say that a_1 and a_2 are in W means that $a_3 = b_i' a_1 b_i$ and $a_4 = b_j' a_2 b_j$ are in $\mathscr{S}$ for some i and j, where the meaning of the b_h is as described above (the appropriate meaning being taken according as σ is the identity or an involution. A computation carries the equation $s' a_1 s = g' a_2 g$ into

$$(b_i^{-1} s g^{-1} b_j)' a_3 (b_i^{-1} s g^{-1} b_j) = a_4.$$

Here s is rational, but may be made integral by multiplying the equation by a rational integer, without affecting the condition that a_4 is in $\mathscr{S}$. The only variable in $b_i^{-1} s g^{-1} b_j$ is then g^{-1}, which in any case has unit norm. Thus $b_i^{-1} s g^{-1} g_j$ is an integral element of A, of fixed norm, which transforms one element of $\mathscr{S}$ into another, and so by Theorem 6.1, it can only have a finite number of values. It has thus been shown that W, as defined above, is an M-domain.

The following is an immediate consequence of the existence of an M-domain:

Theorem 10.4. G, G_Q, G_Z, K, H being as at the beginning of this section, if H contains an M-domain W for G_Z, and H is connected, then G_Z is finitely generated.

Proof. Construct a one-dimension simplicial complex S as follows. The vertices of S are to be the elements of G_Z. According to the second condition in the definition of an M-domain, with $s = 1$, the intersection $W \cap gW$ is non-empty for just finitely many g in G_Z, say $g_1, \ldots, g_N$. Then two vertices g and h of S will be joined by a 1-simplex if and only if $g = hg_i$ or $h = gg_i$ for one of the g_i. It is clear then that $g_1, \ldots, g_N$ will generate G_Z if and only if S is connected. To prove this, let X be the set of all vertices in a connected component of S, and for the purpose of the proof assume W is open (no generality is lost by this; it simply means that all inequalities in the definition of the Siegel domain $\mathscr{S}$ are taken as strict inequalities). Then $XW = \bigcup_{g \in X} gW$ is open in H. If Y consists of the vertices in a component of S different from X, then $XW \cap YW = \emptyset$. For if this intersection were not empty, it would mean that, for some $x \in X$, $y \in Y$, $xW \cap yW \neq \emptyset$, and so $x^{-1}y$ is one of the g_i. That is to say x, y would be joined by a segment in S, contrary to the supposition that X, Y are different components of S. H is thus a disjoint union of open sets of the type XW, which is impossible for connected H. Thus S must have only one connected component, as was to be shown.

It can also be shown that, if K is a maximal compact group in G, G_Z has a finite fundamental set of relations. For S is the 1-skeleton of the nerve N of the covering $\{\gamma W\}$ of

$H(\gamma \in G_Z)$, and each 2-simplex of N yields a relation between the g_i, a 2-simplex corresponding to a triad $\gamma, \gamma', \gamma''$ such that $\gamma W \cap \gamma' W \cap \gamma'' W \neq \emptyset$. To show that these relations form a fundamental set (they are certainly finite in number) it is sufficient to show that N is simply connected, remembering that in the calculation of the fundamental group of a complex, only the simplexes of dimension ≤ 2 need be considered. However, in this situation, this result does hold. For, K being a maximal compact subgroup of G, H is simply connected. Also the covering $\{\gamma W\}$ has a refinement whose nerve N_1 is of the same homotopy types as H, and so is simply connected. And finally, it is hard to see that every closed path on N is the image of a closed path on N_1 under the simplicial map of N_1 onto N, constructed in the usual way.

An important application of the concept of an M-domain for G_Z is to the study of the compactness and measure-finiteness of the quotient G/G_Z. The question of compactness will be taken up here. The following lemmas are for the general situation described at the beginning of this section, while Theorem 10.7 will be concerned with the special situation described in Theorem 10.2. □

Lemma 10.5. Let G, G_Q, G_Z, K, H, W be as at the beginning of this section, G being in particular a Lie group and W being closed. Then H/G_Z is compact if and only if W is compact.

Proof. The points of H/G_Z are orbits in H under the operation of G_Z. Thus if f is the natural mapping of H on H/G_Z, two points a, b of H have the same image if and only if $a = gb$ for some g in G_Z. Then the condition $G_Z W = H$ implies that $f(W) = H/G_Z$. Thus if W is compact so is H/G_Z. On the other hand, the second condition in the definition of W implies that at most finitely many elements of W lie on a given orbit under G_Z. That is to say only finitely many points of W are mapped by f on any given point of H/G_Z, and this finite number is bounded. Since G_Z is discrete, f is an open mapping, and so it easily follows that, if H/G_Z is compact, so is W. □

Lemma 10.6. With the same assumptions as in the last lemma, G/G_Z is compact if and only if H/G_Z is compact.

Proof. There is a continuous mapping F of G/G_Z onto H/G_Z, namely F carries the coset $G_Z a$ onto the double coset $G_Z aK$. Hence if G/G_Z is compact so is H/G_Z. Conversely, since G is a Lie group and H is a closed subgroup, F is a fibre-bundle projection; the base space is compact, and also the fibre (since K is compact), and so G/G_Z is compact as required. □

These two lemmas will now be applied to the situation described in Theorem 10.2. In the notation of that theorem, G now being the connected component of the identity of the automorphism group of an algebra A_R with an involution (possibly trivial), the result to be obtained is:

Theorem 10.7. The necessary and sufficient condition for G/G_Z to be compact is that either (1) $A = k$, $\sigma =$ identity, or (2) $\sigma(a) = u^{-1} a^* u$, where $u^* = u$ and $\underline{x}^* u \underline{x} \neq 0$ for all non-zero $\underline{x} \in k^n$.

The proof will be preceded by the following preliminary lemma:

Lemma 10.8. G/G_Z is non-compact if and only if the set of all a in $\mathscr{S}$ (a Siegel domain $P(A_R)$) satisfying either $\nu(a)$ = some constant element (in the case σ = identity) or $v'(a^*)^{-1}v = a$, $\nu(a) = \nu(v)$ (in the case $\sigma \neq$ identity) is not compact, where v is connected with the u of Theorem 10.7 by $v = b^* ub$, for some integral b.

Proof. By using non-strict inequalities for defining $\mathscr{S}$, it follows that W, defined as at the beginning of this section, will be closed. By Lemmas 10.5, 10.6, G/G_Z will be compact if and only if W is. Now W is a finite union $\bigcup_j\{a : b'_j ab_j \in \mathscr{S}\}$ with the conditions $\nu(a) = 1$ (if σ = identity) or $u'(a^*)^{-1}u = a$, $\nu(a) = \nu(u)$ (if $\sigma \neq$ identity). Hence if W is not compact one of the sets $\{a : b'_j ab_j \in \mathscr{S}\}$, along with the appropriate auxiliary condition, must be non-compact. Dropping the subscript j, this means that a necessary condition for G/G_Z to be non-compact is that there should exist a Siegel domain $\mathscr{S}$ and an integral b such that the set $\{a : b'ab \in \mathscr{S}\}$ with the condition $\nu(a) = 1$ or $u'(a^*)^{-1}u = a$, $\nu(a) = \nu(u)$ should be not compact. The condition is also sufficient. For if it is remembered that the b_j are picked as representatives of a finite collection of cosets in the set of all integral elements with a given bound for the norms, it will be seen that $\mathscr{S}$ can always be enlarged so that any given b is among the b_j. Then take the b such that $\{a : b'ab \in \mathscr{S}\}$ along with the appropriate condition is not compact. It will follow that the M-domain W so constructed for G_Z is not compact, and so G/G_Z is not compact as required. □

To complete the proof of the lemma the element b must be removed from the condition just obtained. To do this, set $a_1 = b'ab$, $v = b^* ub$, and the stated condition follows at once (with a_1 instead of a).

Proof of Theorem 10.7. The sufficiency of the condition will be proved first, the two cases (1) and (2), in the statement of the theorem being considered separately.

Suppose that condition (1) holds. Then according to Lemma 10.8, it must be shown that a Siegel domain $\mathscr{S}$ can be constructed in $P(A_R)$ with the property that the set of $a \in \mathscr{S}$ satisfying $\nu(a) = 1$ is compact. Now in this situation all the matrices are of order 1×1, and so a Siegel domain is simply the set of elements δ of k_R such that $\frac{\delta}{\mathrm{tr}\delta}$ is in some compact set. The condition $\nu(\delta) = 1$ implies $N(\delta) = 1$, and since $N(\delta) \asymp \mathrm{tr}(\delta)$ it follows that δ itself is in a compact set. Thus the set $\{\delta \in \mathscr{S}, \nu(\delta) = 1\}$ is compact, and so G/G_Z is compact.

Now let condition (2) hold. Now it has to be shown that, for some Siegel domain $\mathscr{S}$, the set $S = \{a : a \in \mathscr{S}, v'(a^*)^{-1}v = a, \nu(a) = \nu(v)\}$ is compact. Here $v = b^* ub$, but b need not be mentioned since the conditions $\underline{x}^* v\underline{x} \neq 0$ for any non-zero $\underline{x} \in k^n$ and $v^* = v$ are equivalent to the conditions on u in (2). It can be assumed here that $n \geq 2$, since the treatment for $n = 1$ is exactly as above, with the condition $\nu(a) = 1$ replaced by $\nu(a) = \nu(v)$. Let $\mathscr{S}$ then be a Siegel domain in $P(A_R)$ and assume that S is not compact. Then there is a sequence $\{a_i\}$ of elements of S of which no subsequence converges to a limit in S. Each a_i can be written as $a_i = t'_i d_i t_i$ (using the definition of a Siegel domain). Here t_i is triangular unipotent and contained in a compact set, and so a subsequence of $\{t_i\}$ converges. Assume the subsequence already picked, so that $\{t_i\}$ converges to t. Also $d_i = \mathrm{diag}(\delta_{i1}, \delta_{i2}, \ldots, \delta_{in})$ where $\delta_{i1} \prec \delta_{i2} \prec \cdots \prec \delta_{in}$ and $\frac{\delta_{ij}}{\mathrm{tr}\delta_{ij}}$ is in a compact set for each i, j. Writing $\Delta_i = \mathrm{tr}\delta_{in}$, and again assuming that, if necessary, a subsequence has been picked, it follows that $\frac{\delta_{ij}}{\Delta_i}$ converges to a limit δ_j. Write $d = \mathrm{diag}(\delta_1, \delta_2, \ldots, \delta_n)$.

Thus $\frac{1}{\Delta_i}a_i$ converges to $t'dt$. No subsequence of the sequence $\{\Delta_i\}$ can converge to a finite limit, since if it did a subsequence of $\{a_i\}$ would converge in S. Hence it can be assumed that Δ_i diverges to infinity. Now the condition $v'(a_i^*)^{-1}v = a_i$, or $v' = a_i v^{-1} a_i^*$, holds for each i. Divide by Δ_i^2 and take the limit, and it follows that $0 = av^{-1}a^*$, where $a = t'dt$. It follows at once that $N(a) = 0$, and so at least one of the δ_i is zero. On the other hand they are not all zero since $\frac{\delta_{in}}{\Delta_i}$ has trace equal to 1, and hence so has δ_n. It follows easily that a can be written in partitioned form as $\begin{pmatrix} 0 & 0 \\ 0 & a_1 \end{pmatrix}$ where a_1 is invertible. Then if v^{-1} is similarly partitioned as $\begin{pmatrix} w_1 & w_2 \\ w_3 & w_4 \end{pmatrix}$, the equation $av^{-1}a^* = 0$ means $a_1 w_4 a_1^* = 0$, on $w_4 = 0$. It follows at once that if $\underline{x} \in k^n$ has its first r entries equal to zero, where r is the number of δ_i equal to zero, $\underline{x}^* v^{-1}\underline{x} = 0$, with $\underline{x} \neq 0$. Setting $\underline{x} = v\underline{y}$ this gives $\underline{y}^* v \underline{y} = 0$, $\underline{y} \neq 0$ in k^n, contrary to the assumed condition (2). Hence S is compact as required.

The sufficiency of the conditions in Theorem 10.7 is thus proved, and now the necessity will be established. This will be done by showing that the negation of the given conditions leads to non-compactness of G/G_Z. That is to say it will be shown that G/G_Z fails to be compact if either (a) $A = M_n(k)$, $n \geq 2$, σ = identity, or (b) $\sigma(a) = u^{-1}a^*u$, $u^* = \pm u$, and $\underline{x}^* u \underline{x} = 0$ has a solution for some non-zero $\underline{x}$ in k^n. (Here $n \geq 2$, since k is a division algebra.)

If (a) holds then, according to Lemma 10.8, it has to be shown that for some Siegel domain $\mathscr{S}$, the set $\{a \in \mathscr{S}, \nu(a) = C\}$ where C is a constant, is not compact. Let $a = t'dt$ be any element of $\mathscr{S}$ such that $\nu(a) = C$, t being triangular unipotent and d diagonal, equal to diag$\{\delta_1, \ldots, \delta_n\}$ in the usual way. Then if d_m is diag$\{\frac{1}{m}\delta_1, \delta_2, \ldots, \delta_{n-1}, m\delta_n\}$, it is clear that $t'd_m t$ has no convergent subsequence, and so this set is not compact, as required.

Now suppose that condition (b) holds. This time, using Lemma 10.8 again, it is to be shown that the set $S = \{a : a \in \mathscr{S}, v'(a^*)^{-1}v = a, \nu(a) = \nu(v)\}$ is not compact where $\mathscr{S}$ is a Siegel domain and v is obtained from u by the transformation $v = b^*ub$ for some integral b. The element b will be chosen in such a way that v takes on a specially simple form. In the first place change the basis in k^n (over k) in such a way that a solution of $x^*ux = 0$ is taken as the first basis element; this corresponds to a transformation of u of the form $u \to a^*ua$ into a matrix with top left hand element equal to zero. Then simultaneous operations on the rows and columns of this matrix bring it into the form

$$\left(\begin{array}{cc|ccc} 0 & 1 & 0 & \cdots & 0 \\ \pm 1 & 0 & 0 & \cdots & 0 \\ \hline 0 & 0 & & & \\ \vdots & \vdots & & u_1 & \\ 0 & 0 & & & \end{array}\right)$$

the sign $\pm$ being chosen according as u^* is $+u$ or $-u$. If u_1 can be transformed so that the top left hand element is zero, repeat the process, and so on. After a finite number of steps this gives a matrix of the form $\begin{pmatrix} \begin{matrix} 0 & 1 \\ \pm 1 & 0 \end{matrix} & & & & \\ & \begin{matrix} 0 & 1 \\ \pm 1 & 0 \end{matrix} & & & \\ & & \ddots & & \\ & & & \begin{matrix} 0 & 1 \\ \pm 1 & 0 \end{matrix} & \\ & & & & u_2 \end{pmatrix}$.

If it is not possible to operate on u_2 so that the top left hand element is zero, then simultaneous operations on the rows and columns of u_2 will reduce it to diagonal form. Finally a rearrangement of the rows and columns brings the matrix into the form

$$w = \left(\begin{array}{c|c|c} 0 & 0 & I \\ \hline 0 & \begin{matrix} w_1 & & \\ & \ddots & \\ & & w_r \end{matrix} & 0 \\ \hline \pm I & 0 & 0 \end{array}\right)$$

where $w_i^* = \pm w_i$ and I denotes a unit matrix. This matrix, by its construction is $b_0^* u b_0$, for some $b_0 \in M_n(k)$. Multiplying up by a suitable rational integer to make b_0 integral, it follows that, for an integral matrix b, $b^* ub$ can be written as:

$$v = b^* ub = \left(\begin{array}{c|c|c} 0 & 0 & cI \\ \hline 0 & \begin{matrix} v_1 & & \\ & \ddots & \\ & & v_r \end{matrix} & 0 \\ \hline \pm cI & 0 & 0 \end{array}\right)$$

where c is rational integer. This will be taken as the element v for the application of Lemma 10.8. Let d be a diagonal matrix belonging to the set $S = \{a : a \in \mathscr{S}, v'(a^*)^{-1} v = a, \nu(a) = \nu(v)\}$, and write $d = \mathrm{diag}\{\delta_1, \ldots, \delta_n\}$. The δ_i must satisfy the usual conditions for d to belong to a Siegel domain. The condition $\nu(d) = \nu(v)$ is a bounding condition on the product of the norms of the δ_i. And it is not hard to compute that the condition $v'(d^*)^{-1} v = d$ means that the δ_i can be divided into three sets ($n - r$ must be even) which can be written as $\alpha_1, \ldots, \alpha_s$, $\beta_1, \ldots, \beta_r$, $\gamma_1, \ldots, \gamma_s$, where $v_i'(\beta_i^*)^{-1} v_i = \beta_i$ for each i and $\gamma_i = c^2(\alpha_i^*)^{-1}$ for each i. For any integer m it is clear that diagonal matrix $d_m = \mathrm{diag}\{\frac{1}{m}\alpha_1, \ldots, \frac{1}{m}\alpha_s, \beta_1, \ldots, \beta_r, m\gamma_1, \ldots, m\gamma_s\}$ still satisfies the conditions for lying in S, and the sequence $\{d_m\}$ does not converge to any finite limit. Thus S is not compact as required. The proof of Theorem 10.7 is thus completed. □

5

Fundamental Sets for Arithmetic Groups*

Armand Borel

Translated by Lizhen Ji

Introduction

Let G be a linear semi-simple algebraic group defined over the field $\mathbb{Q}$ of rational numbers. When G is identified with a matrix group, for any subring B of $\mathbb{C}$, we denote, following the custom, by G_B the subgroup of elements of G whose matrix coefficients are contained in B and whose determinants are units in B. In particular $G_{\mathbb{R}}$ is a real semi-simple Lie group, having finitely many connected components, and $G_{\mathbb{Z}}$ is a discrete subgroup, the *group of units* of G. Two matrix realizations of G defined over $\mathbb{Q}$ result in two commensurable groups of units.[1] The results of this paper are essentially concerned with the groups of units; however, it is convenient to consider also subgroups of finite index of groups of units, or what amounts to the same, the subgroups of $G_{\mathbb{Q}}$ commensurable with groups of units, which we call *arithmetic groups* in this paper.[2]

Let K be a maximal compact subgroup of $G_{\mathbb{R}}$, $D = K\backslash G_{\mathbb{R}}$ the Riemannian symmetric space of negative curvature of G, and Γ an arithmetic subgroup of G. We will call a *fundamental set* for Γ in D any subset Ω of D having the following two properties:

(i) $\Omega \cdot \Gamma = D$.
(ii) (*Siegel property*) For every $a \in G_{\mathbb{Q}}$, the set of elements $x \in \Gamma$ for which $\Omega \cdot a \cap \Omega \cdot x \neq \emptyset$ is finite.

* Translated from *Ensembles fondamentaux pour les groupes arithmétiques*, Colloque sur la Théorie des Groupes Algébriques, Bruxelles 1962, pp. 23–40. Librairie Universitaire, Louvain; Gauthier-Villars, Paris.

1 Recall that two subgroups A and B of a group are said to be commensurable if $A \cap B$ is of finite index in each of them.

2 This is for the sake of simplicity, although there are subgroups of finite index of $SL(2,\mathbb{Z})$ which do not contain congruence subgroups of $SL(2,\mathbb{Z})$, so they are not of arithmetic nature in the classical sense.

The existence of open or closed fundamental sets of finite invariant measure, which is classical in a number of particular cases, was established in the general fashion in [3]. Here we propose to describe fundamental sets, in general different from those in [3], which, when D/Γ is not compact, can be considered as union of compact subsets and *cusps* (cf. §2) and study, from the topological point of view, compactifications of D/Γ (cf. §4). We will see that these cusps play a role similar to those of parabolic cusps of fundamental domains of Fuchsian groups. We will need for that, on the one hand, some properties of algebraic groups defined over perfect fields recalled in §1, and on the other hand, some results of Satake on compactifications of D associated with linear representations of G recalled in §3. In §4, we will also show that cusps of fundamental sets are represented by double cosets of $G_{\mathbb{Q}}$ with respect to Γ and $H_{\mathbb{Q}}$, where H is a minimal parabolic subgroup defined over $\mathbb{Q}$, and we will give other interpretations of cusps in some particular cases.

In this Note, proofs are omitted, or at most sketched. They will be published later.

Remark. The condition (ii) is applicable only to arithmetic groups. To make it apply to more general situations, we can either weaken it by simply asking that the set of $x \in \Gamma$ for which $\Omega \cap \Omega \cdot x \neq \emptyset$ is finite, or strengthen it by demanding (ii) for all elements of *a group* $\tilde{\Gamma}$ *of transformation* of $G_{\mathbb{R}}/\Gamma$ [9]. Recall that $\tilde{\Gamma}$ is by definition a subset of elements $a \in G_{\mathbb{R}}$ such that $a \cdot \Gamma \cdot a^{-1}$ is commensurable with Γ; it contains $G_{\mathbb{Q}}$ but can be distinct. However we can show that *if* N *is the largest normal subgroup defined over* $\mathbb{Q}$ *such that* $N_{\mathbb{R}}$ *is compact, then* $\tilde{\Gamma}$ *is the inverse image of* $(G/N)_{\mathbb{Q}}$ *by the canonical projection.* It is easy to deduce that the fundamental sets in [3] or §2.4 still verify thus strengthened condition (ii) (so that arithmetic groups are *Minkowskian* in the terminology of [9]), and several results of this paper are also valid for subgroups of $G_{\mathbb{R}}$ (and not only of $G_{\mathbb{Q}}$) commensurable with the group of units.

1 Algebraic groups over perfect fields[3]

1.1. Let T be an algebraic torus, and k the field of definition of T. We denote by $X(T)$ the group of rational characters of T (rational homomorphisms of T to the multiplicative group of the universal field). We know that $X(T)$ is a free commutative group of rank equal to the dimension T, and the elements of $X(T)$ are all defined over k if and only if T is *split* over k, that is to say, is isomorphic over k to a product of the multiplicative group of the universal field.

1.2. From now on, k is a perfect field. Among all k-split tori of an algebraic group G, the maximal tori are conjugate by elements of G_k. Their common dimension is called the k-rank $r_k(G)$ of G. Let ${}_kT$ be a maximal k-split torus, $\mathscr{N}({}_kT)$ and $\mathscr{Z}({}_kT)$ its normalizer and centralizer in G. The group ${}_kW(G) = \mathscr{N}({}_kT)/\mathscr{Z}({}_kT)$ is the Weyl group of G relative to k. It is finite. The nontrivial characters of ${}_kT$ in the adjoint representation are the

[3] Many of the results of this paragraph are also obtained, independently, sometimes under more general assumptions about fields of definition, by J. Tits [10]. The proofs of some of them, communicated by R. Godement, have been given in the exposition [6]. We assume that the reader is familiar with the theory of linear algebraic groups defined over algebraically closed fields.

roots of G relative to k, or the k-roots of G. The set of roots relative to k is denoted by ${}_k\Sigma(G)$ or ${}_k\Sigma$. Suppose G is connected and semi-simple. Then $X({}_kT)$ is generated, as a vector space over $\mathbb{Q}$, by ${}_k\Sigma$. For each $\alpha \in {}_k\Sigma$, there exists an element $s_\alpha \in {}_kW$, necessarily unique, which transforms α to $-\alpha$ and induces the identity map on $X({}_kT)/\mathbb{Z}\cdot\alpha$, and ${}_kW$ is generated by these s_α. Once an order is chosen on $X({}_kT)$, any k-root is a linear combination with integer coefficients of the same sign of $r_k(G)$ positive roots, which are called simple k-roots, whose set is denoted by ${}_k\Delta$. In particular, the Weyl group relative to k is always crystallographic. There exists a unipotent subgroup ${}_kN$ defined over k, whose normalizer contain ${}_kT$ and its Lie algebra is the sum of eigenspaces of ${}_kT$ correspond to the positive roots, and we have *the Bruhat Lemma* for G_k:

$$G_k = {}_kN_k \cdot \mathcal{N}({}_kT)_k \cdot {}_kN_k, \quad \mathcal{N}({}_kT) = \mathcal{N}({}_kT)_k \cdot \mathcal{Z}({}_kT). \tag{1}$$

1.3. A closed subgroup H of a connected algebraic group is called *parabolic* if G/H is a complete variety. If H is also defined over k, the fibration of G by H possesses local rational sections defined over k, and the projection $G_k \to (G/H)_k$ is surjective. These parabolic subgroups defined over k and conjugate in G are also conjugate by elements of G_k. The minimal parabolic subgroups defined over k are conjugate. A parabolic subgroup is equal to its normalizer.

Suppose G is semi-simple. Minimal parabolic subgroups defined over k are conjugate by the subgroup $H = \mathcal{Z}({}_kT) \cdot {}_kN$ (cf. §1.2), the semidirect product of $\mathcal{Z}({}_kT)$ and the normal subgroup ${}_kN$. Let θ be a subset, possibly empty, of ${}_k\Delta$, and T_θ the identity component of the intersection of the kernel of the characters $\alpha \in \theta$ of ${}_kT$. Then ${}_kN$ and the centralizer $\mathcal{Z}(T_\theta)$ of T_θ generate a parabolic subgroup H_θ defined over k, and every parabolic subgroup defined over k is conjugate to one H_θ and a unique one.

1.4. Until the end of §1, k is of characteristic zero, and G is a connected semi-simple algebraic group defined over k. Then the subgroup ${}_kN$ of §1.2 is maximal among unipotent subgroups defined over k, and any unipotent subgroup defined over k is conjugate, by an element of G_k, to a subgroup of ${}_kN$. The minimal parabolic subgroups defined over k are the normalizers of the maximal unipotent subgroups defined over k. The group $\mathcal{Z}({}_kT)$ contains a unique connected algebraic subgroup ${}_kM$, defined over k, such that $\mathcal{Z}({}_kT) = {}_kM \cdot {}_kT$ and ${}_kM \cap {}_kT$ is finite. The group ${}_kM$ is *anisotropic*, which means ${}_kM_k$ consists of semi-simple elements and ${}_kM$ does not admit any nontrivial rational character defined over k. The k-rank of G is zero if and only if G is anisotropic. The k-roots satisfy the integral condition: if $(,)$ is a positive nondegenerate scalar product on $X({}_kT) \otimes \mathbb{Q}$, invariant under ${}_kW$, then the number $2(\alpha,\beta)/(\alpha,\alpha) = q(\beta,\alpha)$ is an integer for all $\alpha, \beta \in {}_k\Sigma$.

1.5. [4] Let ${}_kT$ be a maximal k-split torus of G, and T a maximal torus of G defined over k which contains ${}_kT$, and k' a finite Galois extension of k on which T splits. Then the

[4] If k is algebraically closed, the k-weights are none other than the weights in the usual sense, and the results of §1.5 and §1.6 are classical. However they are formulated in general in a different way, because here the roots are characters of ${}_QT$, but not linear forms of the Lie algebras, and we consider representations on the right. In the case $k = \mathbb{R}$, the results of §1.5 are due to Satake [7].

Galois group $\mathrm{Gal}(k'/k)$ of k' over k acts on $X(T)$ and on the set of (absolute) roots of G with respect to T. We suppose the orders on $X(T)$ and $X({}_kT)$ are chosen in such a way such that

$$\alpha \geq 0 \Longrightarrow \alpha|_{{}_kT} \geq 0 \quad (\alpha \in X(T)), \tag{1}$$

$$\alpha > 0,\ \alpha|_{{}_kT} \neq 0 \Longrightarrow s(\alpha) > 0\ (\alpha \in X(T), s \in \mathrm{Gal}(k'/k)). \tag{2}$$

If ρ is a linear *representation* of G from the right on a finite dimensional vector space V, we put, for any $\lambda \in X({}_kT)$:

$$V_\lambda = \{v \in V \mid v \cdot \rho(t) = \lambda(t) \cdot v \quad (t \in {}_kT)\}.$$

The character λ is called a k-weight, or a weight relative to k, of ρ if $V_\lambda \neq 0$. Suppose ρ is irreducible, let μ_ρ be the minimal weight with respect to T and the chosen order, and λ_ρ the restriction of μ_ρ to ${}_kT$. Then any k-weight of ρ is of the form

$$\lambda = \lambda_\rho \cdot \alpha_1^{m_1} \cdots \alpha_r^{m_r} \quad (m_i \in \mathbb{Z}; m_i \geq 0\ (1 \geq i \geq r);\ \{\alpha_1, \ldots, \alpha_r\} = {}_k\Delta). \tag{3}$$

Let $c(\lambda)$ be the set of $\alpha_i \in {}_k\Delta$ for which $m_i \neq 0$. Then a subset θ of ${}_k\Delta$ is of the form $c(\lambda)$ if and only if $\theta \cup \lambda_\rho$ is connected, in other words not the union of two nonempty subsets, which are disjoint and mutually orthogonal. For any such subset θ, let $V_\theta = \Sigma_{c(\lambda)\subset\theta} V_\lambda$, and θ' be the set of elements of ${}_k\Delta$ orthogonal to $\theta \cup \lambda_\rho$. Then

$$\mathcal{N}(V_\theta) = \{g \in G \mid V_\theta \cdot \rho(g) = V_\theta\} = H_{\theta\cup\theta'}$$

(in the notation in §1.3). Let $\mathfrak{g}_\alpha$ be the eigensubspace of the Lie algebra $\mathfrak{g}$ of G corresponding to $\alpha \in {}_k\Sigma$. The sum of the subspaces $\mathfrak{g}_\alpha + [\mathfrak{g}_\alpha, \mathfrak{g}_{-\alpha}]$, where α runs through all k-roots which are linear combinations of elements of θ, is the Lie algebra of a semisimple connected linear algebraic subgroup L_θ, defined over k, of k-rank equal to the number of elements of θ, and we have

$$\mathcal{N}(V_\theta) = L_\theta \cdot \mathcal{Z}(V_\theta) \cdot \mathcal{Z}({}_kT), \tag{4}$$

where $\mathcal{Z}(V_\theta)$ is the set of elements of $\mathcal{N}(V_\theta)$ whose restriction to V_θ is a scalar multiple of the identity. $\mathcal{Z}(V_\theta)$ is a normal subgroup whose intersection with L_θ is finite. The group $\mathcal{Z}({}_kT)$ is the normalizer of L_θ; the quotient $\mathcal{N}(V_\theta)/\mathcal{Z}(V_\theta)$ is isogenous to the product of L_θ with a normal subgroup of ${}_kM$; its k-rank is equal to that of L_θ.

1.6. There exist $r = r_k(G)$ rational irreducible representations $\tilde{\omega}_1, \ldots, \tilde{\omega}_r$ of G, defined over k, such that all rational irreducible representations of G, defined over k and the minimal weights (for the order chosen above) defined over k, are those whose minimal weights are linear combinations with integer coefficients ≥ 0 of the minimal weights μ_i of the representations $\tilde{\omega}_i$. We have

$$2(\alpha_i, \lambda_j) = e_j \cdot (\alpha_i, \alpha_j) \cdot \delta_{ij} \quad (1 \leq i, j \leq r; e_j \in \mathbb{Z}, e_j > 0).$$

1.7. Let $k = \mathbb{R}$. Then ${}_kT$ is maximal among all tori of G which are defined over $\mathbb{R}$ and all its real elements have real eigenvalues, and $N = {}_{\mathbb{R}}N_{\mathbb{R}}$ is a maximal unipotent subgroup of $G_{\mathbb{R}}$. Let A be the connected component (in the ordinary topology) containing the identity element of $N = {}_{\mathbb{R}}N_{\mathbb{R}}$, and let K be a maximal compact subgroup of $G_{\mathbb{R}}$ whose Lie algebra $\mathfrak{k}$ is orthogonal to that of A, with respect to the Killing form. Then $G_{\mathbb{R}} = K \cdot A \cdot N$, more precisely the map $(k, a, n) \to k \cdot a \cdot n$ $(k \in K, a \in A, n \in N)$ is an analytic homeomorphism of $K \times A \times N$ on $G_{\mathbb{R}}$ (the *Iwasawa decomposition* of $G_{\mathbb{R}}$). Recall again that if $\mathfrak{p}$ is the orthogonal complement of $\mathfrak{k}$ in $\mathfrak{g}_{\mathbb{R}}$, with respect to the Killing form, then $(k, p) \to k \cdot \exp p$ $(k \in K, p \in \mathfrak{p})$ is a real analytic homeomorphism of $K \times \mathfrak{p}$ on $G_{\mathbb{R}}$ (the *Cartan decomposition*), and $\mathfrak{a}$ is a maximal subalgebra of $\mathfrak{p}$. There exists only one involutive automorphism s of $G_{\mathbb{R}}$, whose K is the set of fixed points; it maps every element x of $P = \exp \mathfrak{p}$ on its inverse; it is the *Cartan involution* of $G_{\mathbb{R}}$ associated to $\mathfrak{k}$.

2 Fundamental sets

2.1. Let G be a semi-simple algebraic group defined over $\mathbb{Q}$ and Γ an arithmetic subgroup of G. A subset $\Omega \subset G_{\mathbb{R}}$ is called *fundamental* for Γ if it verifies the following conditions:

(i′) $\Omega \cdot \Gamma = G$.
(ii′) (*Siegel property*) For every $a \in G_{\mathbb{R}}$, the set of $x \in \Gamma$ such that $\Omega \cdot a \cap \Omega \cdot x \neq \emptyset$ is finite.
(iii′) $K \cdot \Omega = \Omega$ for an maximal compact subgroup K of $G_{\mathbb{R}}$.

Let π be the natural projection of $G_{\mathbb{R}}$ on $D = K \backslash G_{\mathbb{R}}$. It is clear that $\pi(\Omega)$ is a fundamental set for Ω in D in the sense of introduction, and the inverse image of a fundamental set in D is a fundamental set in $G_{\mathbb{R}}$. Furthermore, $G_{\mathbb{R}}$ is a fiber bundle over D with compact fibers; therefore Ω is compact, or closed, or of finite invariant measure if and only if $\pi(\Omega)$ is. It is therefore the same to study fundamental sets in $G_{\mathbb{R}}$ or D. In this paragraph, we consider $G_{\mathbb{R}}$.

2.2. Let ${}_{\mathbb{Q}}T$ be a maximal $\mathbb{Q}$-split torus, H a minimal parabolic subgroup defining over $\mathbb{Q}$ containing ${}_{\mathbb{Q}}T$, and K a maximal compact subgroup of $G_{\mathbb{R}}$ whose Lie algebra is orthogonal, with respect to the Killing form, to that of ${}_{\mathbb{Q}}T_{\mathbb{R}}$ (we will sometimes say K and ${}_{\mathbb{Q}}T$ are adapted to each other). We will take the notations in §1.2, §1.3, and write H in the form $H = {}_{\mathbb{Q}}M \cdot {}_{\mathbb{Q}}T \cdot {}_{\mathbb{Q}}N$, where ${}_{\mathbb{Q}}N$, the unipotent radical of H, is a maximal unipotent subgroup of G defined over $\mathbb{Q}$, and ${}_{\mathbb{Q}}M$ is a largest anisotropic connected subgroup of $\mathcal{Z}({}_{\mathbb{Q}}T)$. We note that ${}_{\mathbb{Q}}\Delta$ is the set of simple $\mathbb{Q}$-roots relative to the order associated with ${}_{\mathbb{Q}}N$, i.e., with respect to which the weights of ${}_{\mathbb{Q}}T$ in the Lie algebra of ${}_{\mathbb{Q}}N$ are positive.

Let ${}_{\mathbb{Q}}A$ be the connected component, in the ordinary topology, of the identity element in ${}_{\mathbb{Q}}T_{\mathbb{R}}$. For each $t \in \mathbb{R}$, we put

$$ {}_{\mathbb{Q}}A_t = \{a \in {}_{\mathbb{Q}}A \mid \alpha(a) \leq e^t, \alpha \in {}_{\mathbb{Q}}\Delta\}. \tag{1} $$

We call a *generalized Siegel domain* in $G_{\mathbb{R}}$ (resp. of $K \backslash G_{\mathbb{R}}$) any subset

$$ \mathfrak{S}_{t,\eta,\omega} = K \cdot \eta \cdot {}_{\mathbb{R}}A_t \cdot \omega, \tag{2} $$

where $t \in \mathbb{R}; \eta \subset {}_{\mathbb{Q}}M_{\mathbb{R}}; \omega \subset {}_{\mathbb{Q}}N_{\mathbb{R}}; \eta, \omega$ are compact (resp. the image of $\mathfrak{S}_{t,\eta,\omega}$ in $K \backslash G_{\mathbb{R}}$) under the canonical projection.

It is immediate that if $h \in H_{\mathbb{R}}$, the set $\mathfrak{S}_{t,\eta,\omega} \cdot h$ is contained in a generalized Siegel domain.

In fact, we should talk about generalized domains relative to ${}_{\mathbb{Q}}T, H, K$, since the proceeding definition presupposes the choice of these three groups, the last being adapted to the first; however we will omit this precision when it will not cause confusion.

2.3. If we will take everything in the above definition *relative to* $\mathbb{R}$, we recover Siegel domains in the usual sense [3, 8]: Let $G_{\mathbb{R}} = K \cdot A \cdot N$ be the Iwasawa decomposition of $G_{\mathbb{R}}$ (cf. §1.7) such that $A \supset A_{\mathbb{Q}}$ and $N \supset {}_{\mathbb{Q}}N_{\mathbb{R}}$, which always exist. A Siegel domain is a set

$$\mathfrak{S}_{t,\omega} = K \cdot A_t \cdot \omega \quad (\omega \text{ compact in } N),$$

where

$$A_t = \{a \in A \mid \alpha(a) \leq e^t, \alpha \in {}_{\mathbb{R}}\Delta\}$$

(cf. [3, §4]). If $A = {}_{\mathbb{Q}}A$, which is equivalent to say that ${}_{\mathbb{Q}}T$ is also maximal $\mathbb{R}$-split torus, then ${}_{\mathbb{Q}}M_{\mathbb{R}} \subset K$, and the generalized Siegel domains are none other than the Siegel domains in the usual sense. In the general case, we can see easily that the Lie algebra ${}_{\mathbb{Q}}\mathfrak{m}_{\mathbb{R}}$ of ${}_{\mathbb{Q}}M_{\mathbb{R}}$ is orthogonal to that of $A_{\mathbb{Q}}$. It follows that $Z_{\mathbb{R}}$ is invariant under the Cartan involution s of $G_{\mathbb{R}}$ whose fixed point set is K, and

$${}_{\mathbb{Q}}M_{\mathbb{R}} = (K \cap {}_{\mathbb{Q}}M) \cdot (A \cap {}_{\mathbb{R}}M) \cdot (N \cap {}_{\mathbb{Q}}M)$$

is the Iwasawa decomposition of ${}_{\mathbb{Q}}M_{\mathbb{R}}$ (cf. [3, §1]). Since η is compact, and ${}_{\mathbb{Q}}M$ centralizes ${}_{\mathbb{R}}T$, we can immediately deduce that $\mathfrak{S}_{t,\eta,\omega}$ is contained in a Siegel domain of $G_{\mathbb{R}}$ relative to the decomposition $K \cdot A \cdot N$. The set $\mathfrak{S}_{t,\eta,\omega}$ is therefore of *finite Haar measure* [3, §4.3].

Notice that, as for Siegel domains in the usual sense [3, §4.1], we could in §2.2 (1) let α go through all positive roots, or replace ${}_{\mathbb{Q}}A_t$ by $a \cdot {}_{\mathbb{Q}}A_0 (a \in {}_{\mathbb{Q}}A)$ without changing anything essential.

2.4. Theorem. *Let G be a semi-simple algebraic group defined over $\mathbb{Q}$ and Γ an arithmetic subgroup of G. Then there exist a generalized Siegel domain $\mathfrak{S}_{t,\eta,\omega}$ of $G_{\mathbb{R}}$ and a finite number of elements $x_i \in G_{\mathbb{Q}}$ $(1 \leq i \leq m)$ such that*

$$\Omega = \bigcup_{i=1}^{m} \mathfrak{S}_{t,\eta,\omega} \cdot x_i \quad (x_i \in G; i = 1, \ldots, m) \tag{A}$$

is a fundamental set for Γ in $G_{\mathbb{R}}$. Any subset in the form (A) *is of finite Haar measure and possesses the Siegel property.*

The proof takes as the point of departure of the existence of fundamental set of the type indicated in [3, Theorem 6.5] and the validity of conjecture of Godement [3, Theorem 11.6]. It uses several results mentioned in §1 and parts of reasonings analogous to those in §7 of [3].

Note that ${}_{\mathbb{Q}}A_{s,t}$ $(s < t)$ is the set of elements of ${}_{\mathbb{Q}}A$ on which the simple $\mathbb{Q}$-roots take values between e^s and e^t. Then Ω is the union of a compact subset $\cup K \cdot \eta \cdot {}_{\mathbb{Q}}A_{s,t} \cdot \omega \cdot x_i$ and the sets $\mathfrak{S}_{s,\eta,\omega} \cdot x_i$. It is these latter sets which we call the *cusps* of Ω, with the understanding that s takes a negative value, whose absolute value is sufficient large.

2.5. Example. Let G be the orthogonal group of a rational quadratic form, non-degenerate and isotropic, which we suppose its matrix F can be put in the form

$$F = \begin{pmatrix} 0 & 0 & S \\ 0 & F_0 & 0 \\ S & 0 & 0 \end{pmatrix}, \tag{1}$$

where

$$S = \begin{pmatrix} 0 & & 1 \\ & \cdot^{\cdot^{\cdot}} & \\ 1 & & 0 \end{pmatrix},$$

and F_0 does not represent zero rationally. The degree of S is therefore the dimension of the maximal isotropic subspace of F which is defined over $\mathbb{Q}$. We can also take for H the set of matrices of G which have the form

$$X = \begin{pmatrix} A_1 & A_2 & A_3 \\ 0 & B_2 & B_3 \\ 0 & 0 & C_3 \end{pmatrix}, \tag{2}$$

with A_1 upper triangular. It is elementary, and well-known, that A_1, A_2 are arbitrary, and C_3 is upper triangular and determined by A_1, B_2 runs through the orthogonal group $O(F_0)$ of F_0, and A_1, A_2, B_2 also determine A_3 and B_3. The group H in none other than the stability group of the maximal flag of isotropic subspaces $E_1 \subset \cdots \subset E_d$, where E_i is the isotropic subspace spanned by the first i basis vectors. The minimal parabolic subgroups defined over $\mathbb{Q}$ are the stability groups of the maximal flags of isotropic subspaces defined over $\mathbb{Q}$. We can take for ${}_{\mathbb{Q}}T$ the set of diagonal matrices of H for which

$$B_2 = Id., \quad A_1 = \begin{pmatrix} x_1 & & & 0 \\ & x_2 & & \\ & & \ddots & \\ 0 & & & x_d \end{pmatrix}, \quad C_3 = \begin{pmatrix} x_d^{-1} & & & 0 \\ & x_{d-1}^{-1} & & \\ & & \ddots & \\ 0 & & & x_1^{-1} \end{pmatrix}. \tag{3}$$

Then we have

$$\mathcal{Z}({}_{\mathbb{Q}}T) = {}_{\mathbb{Q}}T \times {}_{\mathbb{Q}}M, \quad {}_{\mathbb{Q}}M = O(F_0). \tag{4}$$

Let $\mathbb{Q}$-roots are the characters $x_i \cdot x_j^{\pm 1}$ and their inverses $(1 \le i < j \le d)$, to which we must add also $x_i^{\pm 1}$ $(1 \le i \le d)$ when $F_0 \ne 0$. For the simple $\mathbb{Q}$-roots relative to the order defined by H, and for the definition of ${}_{\mathbb{Q}}A_t$, we have two distinguished cases:

(i) $F_0 = 0$. The simple $\mathbb{Q}$-roots are

$$x_i x_{i+1}^{-1} \ (1 \le i \le d-1), \quad x_{d-1} \cdot x_d,$$

and we have

$${}_{\mathbb{Q}}A_t = \{a \in {}_{\mathbb{Q}}A \mid x_i \le e^t \cdot x_{i+1} \ (1 \le i \le d-1);\ x_{d-1} \cdot x_d \le e^t\}.$$

(ii) $F_0 \ne 0$. The simple $\mathbb{Q}$-roots are

$$x_i \cdot x_{i+1}^{-1} \ (1 \le i \le d-1), \quad x_d,$$

and we have

$$_{\mathbb{Q}}A_t = \{a \in {_{\mathbb{Q}}A} \mid x_i \le e^t \cdot x_{i+1} \ (1 \le i \le d-1);\ x_d \le e^t\}.$$

For K, we have to take the intersection of G with the orthogonal group of the majorant of F (in the sense of Hermite), whose Lie algebra is orthogonal to that of $_{\mathbb{Q}}A$. Write the vector space V_0 of F_0 as direct sum of two orthogonal subspaces V_1, V_2 such that the restriction F_i of F_0 to V_i is positive for $i = 1$, and negative for $i = 2$. Then we can take K as the product of the orthogonal group of $-F_2$ with the orthogonal group of the sum of F_1 on V_1 with the unit form in the space of the first d and last d coordinates. If F_0 is positive or negative, then $_{\mathbb{Q}}M_{\mathbb{R}}$ is contained in K, $_{\mathbb{Q}}T$ is also a maximal $\mathbb{R}$-split torus, and the generalized Siegel domains coincide with the Siegel domains in the usual sense (cf. §2.3).

3 Boundary components of symmetric spaces

In all of this paragraph, G is a semi-simple algebraic group defined over $\mathbb{R}$, $G_{\mathbb{R}} = K \cdot P$ is the Cartan decomposition of $G_{\mathbb{R}}$ and s is the corresponding Cartan involution (cf. §1.7). K is thus a maximal compact subgroup, $P = \exp \mathfrak{p}$, where $\mathfrak{p}$ is the orthogonal complement of the Lie algebra $\mathfrak{k}$ of K inside $\mathfrak{g}_{\mathbb{R}}$, relative to the Killing form, and $s(x) = x$ if $x \in K$, $s(x) = x^{-1}$ if $x \in P$. We denote by A the identity component, for the ordinary topology, of the set of real points of an maximal $\mathbb{R}$-split torus, and we suppose, what is permissible, that $A \subset P$. We choose an order on the $\mathbb{R}$-roots and denote by N the real part of the unipotent subgroup corresponding to positive roots.

3.1. Let ρ be an irreducible locally faithful representation of $G_{\mathbb{R}}$ from the right on a vector space (real or complex) V of finite dimension. We suppose that V is given a Hilbert structure such that $\rho(x)$ is unitary (resp. self-adjoint positive) if $x \in K$ (resp. $x \in P$), which it is always possible, and we denote by a^* the adjoint of an endomorphism a of V. Let $\mathscr{E}\mathscr{S}(V)$ be the vector space of self-adjoint endomorphisms of V, and $P\mathscr{E}\mathscr{S}(V)$ the corresponding projective space. The group $G_{\mathbb{R}}$ acts on $\mathscr{E}\mathscr{S}(V)$ and $P\mathscr{E}\mathscr{S}(V)$ by the map: $a \mapsto a \cdot \rho_0(g) = \rho(g)^* \cdot a \cdot \rho(g)$. We will also write sometimes $a \cdot g$ instead of $a \cdot \rho_0(g)$. Since $\rho(P)$ consists of positive definite self-adjoint endomorphisms of determinant 1, the map $g \mapsto \rho(g)^* \cdot \rho(g)$ defines an equivariant embedding of $D = K\backslash G_{\mathbb{R}}$ in $P\mathscr{E}\mathscr{S}(V)$, which we also denote by ρ_0. The closure $\overline{\rho_0(D)}$ of D, which is denoted by D_ρ, is the *compactification of D associated to ρ*. It is thus a transformation space of $G_{\mathbb{R}}$. Satake [7] showed that $D_\rho - D$ is union of finite number of orbits of $G_{\mathbb{R}}$. Furthermore, each of these orbits is a fiber space in symmetric spaces, embedded in linear subspaces of $P\mathscr{E}\mathscr{S}(V)$ as D is in $P\mathscr{E}\mathscr{S}(V)$[5], called the boundary components of D_ρ. It is also common to consider D as a (improper) component of D_ρ. To be complete, we are going to recall description of boundary components.

3.2. We suppose that $\rho(A)$ is put in the diagonal form. The diagonal terms of $\rho(a)$ ($a \in A$) are thus in the form §1.5 (3) (with $k = \mathbb{R}$). Every boundary point is equivalent

[5] However, these embeddings can be defined from non-irreducible representations, which are, but in any case, sums of equivalent irreducible representations.

modulo K to an point of $\overline{\rho_0(A)}$, or even, if K contains representatives of the relative Weyl group ${}_{\mathbb{R}}W$, to a point of $\overline{\rho_0(A^-)}$, where A^- denotes the negative Weyl chamber (the set of points of A on which the positive roots take values ≤ 1). Let (a_n) be a divergent sequence of points in A^-. We are only interested in the limit of $\rho(a_n)^*$. For $\rho(a_n)$ in $P\mathscr{E}\mathscr{S}(V)$, we can in §1.5 (3) ignore the common factor λ_ρ; the limit of $\rho(a_n)^* \cdot \rho(a_n)$ will have its coordinates in V_λ zero if and only if there exists $\alpha \in (\lambda)$ such that $\lim \alpha(a_n) = 0$. From this, and results of §1.5, we can deduce easily the following:

Let Ψ_ρ be the collection of all subsets of ${}_{\mathbb{R}}\Delta$ which together with λ_ρ form connected subsets. Then $\overline{\rho_0(A^-)}$ is contained in the union of the spaces $P\mathscr{E}\mathscr{S}(V_\psi)$ $(\psi \in \Psi_\rho)$. The image of $\mathscr{N}(V_\psi)_{\mathbb{R}}/\mathscr{Z}(V_\psi)_{\mathbb{R}}$ in $P\mathscr{E}\mathscr{S}(V_\psi)$ by the map $x \mapsto \rho(x)^* \cdot \rho(x)$ is the symmetric space F_ψ with negative curvature of $L_{\psi,\mathbb{R}}$. The components of D_ρ are transformations by $G_{\mathbb{R}}$ of the spaces F_ψ $(\psi \in \Psi_\rho)$. If ψ is the greatest element of Ψ_ρ with respect to the inclusion relation, then $V_\psi = V$, and $F_\psi = D$. If ψ is empty, then $V_\psi = V_{\lambda_\rho}$ and L_ψ is reduced to (e); the component F_ψ is a point. If $\psi \subset \psi'$, then F_ψ is a part of the closure of $F_{\psi'}$.

3.3. Let F be a component of D. We put

$$\begin{aligned} \mathscr{N}(F) &= \{g \in G_{\mathbb{R}} \mid F \cdot \rho_0(g) = F\}, \\ \mathscr{Z}(F) &= \{g \in \mathscr{N}(F) \mid f \cdot \rho_0(g) = f \quad (f \in F)\}, \end{aligned} \tag{1}$$

and $G(F) = \mathscr{N}(F)/\mathscr{Z}(F)$. The groups $\mathscr{N}(F)$ and $\mathscr{Z}(F)$ are then conjugates of the groups $\mathscr{N}(V_\psi)_{\mathbb{R}}$ and $\mathscr{Z}(V_\psi)_{\mathbb{R}}$ respectively $(\psi \in \Psi_\rho)$. The group $G(F)$ is isogenous to L_ψ. We have furthermore

$$\mathscr{N}(F) = \{g \in G_{\mathbb{R}} \mid F \cdot \rho_0(g) \cap F \neq \emptyset\}. \tag{2}$$

Two components of D_ρ are said to be the same type if they are part of the same orbit of $G_{\mathbb{R}}$. The group K acts transitively on the set of components of a given type F, which corresponds bijectively to points of $G_{\mathbb{R}}/\mathscr{N}(F)$.

4 Compactifications of D/Γ

In this paragraph, G is a connected semi-simple algebraic group defined over $\mathbb{Q}$, Γ an arithmetic subgroup of G, K a maximal compact subgroup of $G_{\mathbb{R}}$, and $D = K\backslash G_{\mathbb{R}}$.

4.1. Definition. Let ρ be an irreducible representation of $G_{\mathbb{R}}$. A component F of D_ρ (cf. §3.2) is *rational* if its normalizer $\mathscr{N}(F)$ and its centralizer $\mathscr{Z}(F)$ (cf. §3.3) are real algebraic subgroups of $G_{\mathbb{R}}$ defined on $\mathbb{Q}$.

4.2. Lemma. *Let F be a rational component of D_ρ. Then the image $\Gamma(F)$ of $\mathscr{N}(F) \cap \Gamma$ in $G(F)$ by the canonical projection is an arithmetic subgroup.*

The quotient $\mathscr{N}(F)\mathscr{Z}(F)$ is semi-simple, the group $\mathscr{N}(F)$ contains a semi-simple subgroup L defined over $\mathbb{Q}$ such that $L \cap \mathscr{Z}(F)$ is finite and $\mathscr{N}(F) = L \cdot \mathscr{Z}(F)$. The restriction of $\pi : \mathscr{N}(F) \to G(F)$ to L is then an isogeny defined over $\mathbb{Q}$. On the other hand, it follows from [3, §6.3, §6.11] that $\pi(L_{\mathbb{Z}})$ is an arithmetic subgroup of $G(F)$, and $\mathscr{N}(F)_{\mathbb{Z}}$ is commensurable with $L_{\mathbb{Z}} \cdot \mathscr{Z}(F)_{\mathbb{Z}}$, hence the lemma.

4.3. Theorem. *Let ρ be a locally faithful irreducible rational representation of G defined over $\mathbb{Q}$, and D_ρ the compactification of D associated with ρ (§3.1). Let Ω be a properly chosen fundamental set for Γ in D. Then $\overline{\rho_0(\Omega)} = \Omega_\rho$ is contained in the union of finite number of rational components of D_ρ. The quotient $S_\rho = \Omega_\rho \cdot \Gamma/\Gamma$ is the union of D/Γ and a finite number of quotients $F_i/\Gamma(F_i)$, where F_i are rational boundary components; it admits the structure of compact Hausdorff topological space inducing the natural topology on D/Γ and $F_i/\Gamma(F_i)$, such that D/Γ is open in S_ρ and the projection $\Omega_\rho \cdot \Gamma \to S_\rho$ induces a continuous map from Ω_ρ to S_ρ.*

Sketch of proof[6]: We take notations of §1.5, §2.2. λ_ρ is thus the minimal $\mathbb{Q}$-weight of ρ, and Θ_ρ is the collection of all subsets of ${}_{\mathbb{Q}}\Delta$ which form with λ_ρ connected subsets. To study the components of D_ρ, we can use the Iwasawa decomposition $G_{\mathbb{R}} = K \cdot A \cdot N$ of $G_{\mathbb{R}}$ such that $A \supset {}_{\mathbb{Q}}A$ and $N \supset {}_{\mathbb{Q}}N_{\mathbb{R}}$. Then the restriction to ${}_{\mathbb{R}}A$ of a positive $\mathbb{R}$-root (for the order defined by N) is ≥ 0, and ${}_{\mathbb{Q}}A^- \subset A^-$. We conclude easily that the space V_θ (§1.5) can also be written as V_ψ where $\psi \subset {}_{\mathbb{R}}\Delta$ forms with the minimal $\mathbb{R}$-weight of ρ a connected subset. This means, see §3.2, that the image F_θ of $\mathcal{N}(V_\theta)_{\mathbb{R}}/\mathcal{Z}(V_\theta)_{\mathbb{R}}$ in $P\mathcal{E}\mathcal{S}(V_\theta)$ by ρ_0 is a component of D_ρ. Since we can assume that ρ is defined over $\mathbb{Q}$, the space V_θ and the groups $N(V_\theta), Z(V_\theta)$ are also defined over $\mathbb{Q}$, F_θ is rational component, and $\Gamma(F_\theta)$ is an arithmetic subgroup of $G(F_\theta)$.

Let $\mathfrak{S}$ be generalized Siegel domain of D, defined relatively to $K, {}_{\mathbb{Q}}T, H$ (cf. §2.2). Supposing $\rho({}_{\mathbb{Q}}T)$ is put into diagonal form and using §1.5, we can see without difficulty

$$\overline{\rho}_0(\mathfrak{S}) \subset \bigcup_{\theta \in \Theta_\rho} F_\theta, \tag{1}$$

that $\overline{\rho}_0(\mathfrak{S}) \cap F_\theta$ $(\theta \in \Theta_\rho)$ is a generalized Siegel domain of F_θ, defined relative to images of $K \cap \mathcal{N}(V_\theta)$, ${}_{\mathbb{Q}}T$ and H by the natural projection, and any subset of F_θ of this nature is in the closure of image of a generalized domain of D. Let

$$\Omega = \bigcup_{i=1}^{m} \mathfrak{S} \cdot x_i \quad (x_i \in G_{\mathbb{Q}})$$

be a fundamental set for Γ in (cf. §2.4), $\mathfrak{S}$ being thus sufficiently large. The inclusion (1) leads to

$$\Omega_\rho = \overline{\rho_0(\Omega)} \subset \bigcup_{1 \leq i \leq m, \theta \in \Theta_\rho} F_\theta \cdot \rho_0(x_i), \tag{2}$$

which establishes the first assertion.

Applying §2.4 to automorphism group of $F_\theta \cdot \rho_0(x_i)$, we then show that if Ω is sufficiently large, we have

$$F_\theta \cdot \Gamma = (F_\theta \cap \overline{\rho_0(\Omega)}) \cdot \Gamma. \tag{3}$$

Since $F_\theta \cap \mathfrak{S}$ has the Siegel property, we conclude from that the existence of finite number of elements γ_j $(j \in J)$ of Γ having the following property: If $\gamma \in \Gamma$ is such that $\Omega_\rho \cdot \rho_0(\gamma) \cap \Omega_\rho \neq \emptyset$, then there exists $j \in J$ such that $x \cdot \rho_0(\gamma) = x \cdot \rho_0(\gamma_j)$ for all $x \in \Omega_\rho \cap \Omega_\rho \cdot \rho_0(\gamma^{-1})$. We then see that we are in the condition to apply Theorem 1′ of [8], which leads to the second assertion.

[6] Strongly inspired by [8].

It is clear that $\cup_i F_\theta \cdot \rho_0(x_i \cdot \Gamma)$ consists of rational components of the type F_θ. On the other hand, if $x \in G_{\mathbb{Q}}$, then, according to the Siegel property, $\Omega \cdot x$ is contained in union of finite number of translates of Ω by elements of Γ. Therefore, $\Omega_\rho \cdot \rho_0(x) \subset \Omega_\rho \cdot \rho_0(\Gamma)$, from which we deduce $\cup_i F_\theta \cdot \rho(x_i \cdot \Gamma)$ is the union of *all* rational components of type F_θ, and our third assertion.

4.4. Theorem. *Let ${}_{\mathbb{Q}}T$ be a maximal $\mathbb{Q}$-split torus of G, H a minimal parabolic subgroup defined over $\mathbb{Q}$ containing ${}_{\mathbb{Q}}T$, and suppose K is adapted to T. Let $\mathfrak{S}_{t,\eta,\omega}$ be a generalized Siegel domain of $G_{\mathbb{R}}$. The finite union $\Omega = \cup_{1\le i\le m} \mathfrak{S}_{t,\eta,\omega} \cdot x_i$ of translates of $\mathfrak{S}_{t,\eta,\omega}$ by elements $x_i \in G_{\mathbb{Q}}$ is a fundamental set for Γ, for t,η,ω sufficiently large, if and only if the set x_i contains at least representatives of the double cosets $H_{\mathbb{Q}} \backslash G_{\mathbb{Q}} / \Gamma$.*

This statement implies in particular that the number of double cosets $H_{\mathbb{Q}} \backslash G_{\mathbb{Q}} / \Gamma$ is finite, and is equal to the lower bound of the number of cusps of a fundamental set, of type §2.4, for Γ in $G_{\mathbb{R}}$.

For the proof, we choose first of all an irreducible locally faithful representation ρ of G defined over $\mathbb{Q}$, whose minimal $\mathbb{Q}$-weight is not orthogonal to any simple $\mathbb{Q}$-root. The existence of ρ follows from §1.6. By §1.5, if θ is an empty set, V_θ is the weight space corresponding to λ_ρ, $\mathcal{N}(V_\theta) = H$, and $\mathcal{N}(V_\theta) \subset {}_{\mathbb{Q}}N$. There exists thus a rational component F of D_ρ whose normalizer is $H_{\mathbb{R}}$. We then see, as at the end of the proof of Theorem 4.4, that if Ω is a sufficiently large fundamental set, $\cup_i F \cdot \rho_0(x_i \cdot \Gamma)$ is the union of all rational components of type F, and immediately

$$G_{\mathbb{Q}} = \bigcup_{1\le i\le m} H_{\mathbb{Q}} \cdot x_i \cdot \Gamma, \tag{1}$$

which shows that $\{x_i\}$ contains at least one representative of each of the double class of $G_{\mathbb{Q}}$ modulo $H_{\mathbb{Q}}$ and Γ.

Conversely, let $\{y_j\}_{1\le j\le q}$ be a system of representatives of the double cosets. We can write

$$x_i = h_i \cdot y_{j(i)} \cdot \gamma_i \quad (h_i \in H_{\mathbb{Q}}, \gamma_i \in \Gamma; i = 1, \dots, m).$$

Since $\mathfrak{S}_{t,\eta,\omega} \cdot h$ is a part of a generalized Siegel domain for all $h \in H_{\mathbb{R}}$, we see that

$$\bigcup_{1\le i\le m} \mathfrak{S}_{t,\eta,\omega} \cdot x_i \subset \bigcup_{1\le j\le q} \mathfrak{S}_{t',\eta',\omega'} \cdot y_j \cdot \Gamma, \tag{2}$$

for sufficiently large t', η', ω', which implies the theorem.

4.5. Corollary. *Let P be a parabolic subgroup of G defined over $\mathbb{Q}$. Then the rational points of G/P form finite orbits of Γ.*

Let H be a minimal parabolic subgroup defined over $\mathbb{Q}$ and contained in P. The canonical map $(G/H)_{\mathbb{Q}} \to (G/P)_{\mathbb{Q}}$ is surjective, since $G_{\mathbb{Q}} \to (G/P)_{\mathbb{Q}}$ is (§1.3). We are thus back to the case $H = P$, which has been treated in §4.4.

4.6. *Remarks.* (1) By using §1.6. we see easily that Corollary 4.5 is equivalent to the following assertion: Let $\rho : G \to GL(V)$ be a rational representation of G defined over $\mathbb{Q}$. Let L be a lattice of $V_{\mathbb{Q}}$ invariant by Γ and X is a cone of V such that $X - 0$ is one orbit of G. Then the primitive vectors of $L \cap X$ form finite number of orbits of Γ.

Take in particular for G the orthogonal groups of a nondegenerate quadratic form, for Γ the group of units, and for X the isotropic cone. $X - 0$ is one orbit of G according to a theorem of Witt. We thus recover the classical fact that primitive isotropic vectors form a finite number of equivalence classes modulo the group of units of F.

(2) R. Godement told me a direct proof of Corollary 4.5, based on the fact that the adelic group G_A attached to G satisfies the condition (F) (cf. [2]).

4.7. By using several properties of lattices [5, §9, §12], or some adelic groups attached to G and H, we can in some cases give other expressions of number of elements of $H_{\mathbb{Q}}\backslash G_{\mathbb{Q}}/\Gamma$, which we denote it by $\nu(G,\Gamma)$, and which is also, according to §4.5, the minimal number of cusps of a fundamental set. We will confine ourselves to the following examples without proof.

(a) Let $G = O(F)$, where F is a quadratic form §2.5 (1), but we suppose furthermore than the integer points form a maximal lattice for F. Then $\nu(G, G_{\mathbb{Z}})$ is equal to the number of equivalence classes of maximal isotropic planes defined over $\mathbb{Q}$ modulo $G_{\mathbb{Z}}$, or the number of classes in the genre of F_0.

(b) Let k be a number field, h the number of ideal classes of k. Suppose that $G = R_{k/\mathbb{Q}}G'$ is obtained by restriction of scalars from k to $\mathbb{Q}$ in the sense of [11, Chap. 1] from $G' = SL_{n+1}$, $G' = Sp_{2n}$, viewed as groups defined over k. Then $\nu(G, G_{\mathbb{Z}}) = h^n$. If $k = \mathbb{Q}$, we have a unique cusp, a well-known result, given the fact that the generalized Siegel domains are Siegel domains in the usual sense. If k is totally real and $G' = SL_2$, then the group $G_{\mathbb{Z}}$ is the Hilbert-Blumenthal modular group, and we recover the analogue of classical result of H. Maass (Sitzungsberichte Heidelberg. Akad. Wiss., Math-naturwiss. Klasse 1940, 2. Abhandlung).

(c) More generally than in (b), suppose that $G = R_{k/\mathbb{Q}}G'$, where G' is a group of normal (or *Chevalley*) type over k, that is to say it possesses a maximal torus split over k. Then it is practically certain that we have $\nu(G, G_{\mathbb{Z}}) \leq h^r$, and $\nu(G, G_{\mathbb{Z}}) = h^r$ if G' is simply connected,[7] r being the rank of G'. If $k = \mathbb{Q}$, this means that a suitable Siegel domain is a fundamental set, a result proved previously in a very different way by Harish-Chandra, as a part of the Note on automorphic forms (Proc. Nat. Ac. Sci. U.S.A. 45 (1959), 570–573).

Bibliography

[1] A. Borel, Groupes linéaires algébriques, *Ann. of Math.* 64 (1956), 20–80.

[2] A. Borel, Some properties of adele groups attached to algebraic groups, *Bull. Amer. Math. Soc.* 67 (1961) 583–585.

[3] A. Borel, Harish-Chandra, Arithmetic subgroups of algebraic groups, *Ann. of Math.* 75 (1962), 485–535.

[4] C. Chevalley, Séminaire sur la classification des groupes de Lie algébriques, Paris, 1958.

[7] It is probably sufficient to start from a matrix representation of G' in which the lattice of points with coordinates in integers of k is admissible in the sense of Chevalley (Sém. Bourbaki, Exp. 219, 1961). In fact, it would suffice here to know that the representations of SL_2 in G' defined by the radical [root] groups are given by polynomials with integer coefficients in the coefficients of the elements of SL_2.

[5] M. Eichler, Quadratische Formen und orthogonale Gruppen, Grundlehren d.m. Wiss. LXIII, Springer, 1952.
[6] R. Godement, Groupes linéaires algébriques sur un corps parfait, Séminaire Bourbaki, Exp. 206, Décembre 1960.
[7] I. Satake, On representations and compactifications of symmetric Riemannian spaces, *Ann. of Math.* 71 (1960) 77–110.
[8] I. Satake, On compactifications of the quotient spaces for arithmetically defined discontinuous groups, *ibid.* 72 (1960) 555–580.
[9] Séminaire H. Cartan 1957-58. Exposé II par A. Weil.
[10] J. Tits, Groupes semi-simples isotropes, ce Recueil, p. 137.
[11] A. Weil, Adeles and algebraic groups. Notes, The Institute for Advanced Study, Princeton, 1961. (Later published with appendices by M. Demazure and Takashi Ono, Progress in Mathematics, 23. Birkhuser, Boston, Mass., 1982. iii+126 pp.)

6

Fundamental Domains of Arithmetic Groups*

Roger Godement
Translated by Enrico Leuzinger

The fact that one can construct a fundamental domain of finite volume for the group $SL(n,\mathbb{Z})$ in $SL(n,\mathbb{R})$ with the help of simple inequalities has recently been generalized by BOREL and HARISH-CHANDRA to all linear algebraic groups defined over the field of rational numbers (finite volume and compactness, if it holds, are obtained in [2], but the more precise results, which are new even for most classical groups, can be found in [1], a collection of results whose proofs did not yet appear). The compactness criterion for $G_{\mathbb{R}}/G_{\mathbb{Z}}$ on the other hand had been directly proved by MOSTOW and TAMAGAWA [4] with the help of methods completely different from those of BOREL and HARISH-CHANDRA, clearly more simple but which, up to now, had not been extended to the general case.

The goal of this report is to show that one can in fact extend the method of MOSTOW and TAMAGAWA in a way to obtain in the general case the Minkowski inequalities announced by BOREL in [1] and thus obtain a very simple proof of these, even simpler, we hope, than the one of BOREL...This new proof has recently been obtained by A. WEIL and the conferencier.

We assume that the reader knows, besides of course the material on algebraic groups which one can find in the Séminaire Chevalley, in [3], and in the previous, present and future articles of BOREL, only the most basic notions on adelic groups which one will find on the first pages of [6]. If G denotes a linear algebraic group defined over the field $k = \mathbb{Q}$ of rational numbers, one denotes by G_A the corresponding adelic group, and by G_A° the closed subgroup of G_A given by the set of g's such that $|\chi(g)| = 1$ for every rational character χ of G defined over k; one obviously has $G_k \subset G_A^\circ$ by the product formula.

* Translated from *Domaines fondamentaux des groupes arithmétiques*, Séminaire N. Bourbaki, 1962/63, exp. no. 257, pp. 201–225.

1 Minkowski reduction in $GL(n)$

1.1 Heights

Let V be a finite-dimensional vector space defined over k as an algebraic variety (such that for every extension k' of k, the set $V_{k'}$ is obtained from the vector space V_k by extension of scalars to k'). It is clear that V_A is the A-module derived from V_k by extension to A of the base ring k, and that $GL(V)_A$ is canonically identified with $GL(V_A)$. We will say that an $x \in V_A$ is *primitiv* if there exists a $g \in GL(V_A)$ such that $g(x)$ is a non-zero element of V_k. If one has chosen a basis (a_i) of V_k over k, this means that $x_p \neq 0$ for every p and that, for almost every finite p, the coordinates of x_p with respect to this basis are p-adic integers prime among them.

On the set of primitive elements of V_A one calls *height* every function $\|x\|$ with strictly positive values obtained in the following way: for every p (finite or not) one chooses on $V_p = V_{k_p}$ a norm $\|\ \|_p$ compatible with the absolute value defined on k_p, and this in such a way that, for almost every finite p and every $x \in V_p$, the number $\|x\|_p$ is the maximum of the absolute values of the coordinates of x with respect to a basis of V_k chosen once for all. This done, one sets

$$\|x\| = \prod \|x_p\|_p$$

for every primitiv $x \in V_A$. One immediately verifies the following properties:

(i) *if $\|x\|'$ and $\|x\|''$ are heights, the ratio $\|x\|'/\|x\|''$ remains in a fixed compact subset of $\mathbb{R}_+^*$ if x varies,*

(ii) *if primitive elements $x_n \in V_A$ go to 0 in V_A then the $\|x_n\|$ go to 0,*

(iii) *conversely, if the $\|x_n\|$ go to 0, there are non-zero rationals λ_n such that the $\lambda_n x_n$ go to 0 in V_A,*

(iv) *for every $g \in GL(V_A)$ the set of non-zero $\xi \in V_k$ such that $\|g(\xi)\| < c$ is finite modulo k^* (and in particular there is a $\xi \neq 0$ such that $\|g(\xi)\|$ is minimal),*

(v) *one has $\|tx\| = |t| \cdot \|x\|$ for every $t \in A^*$ and every primitiv $x \in V_A$,*

(vi) *if M is a compact part of $GL(V_A)$ there are constants $c', c'' > 0$ such that one has*

$$c'\|x\| \leq \|m(x)\| \leq c''\|x\|$$

for every $m \in M$ and every primitiv $x \in V_A$.

1.2 Minkowski reduction

Let us take for the V above the canonical vector space of dimension n, such that $V_k = k^n$, $V_A = A^n$ and $GL(V) = GL(n)$. In the following we will use the *canonical* height on the set of primitive elements of A^n which is obtained as follows: for p finite and $x \in V_p$ we set $\|x\|_p$ = maximum of the absolute values of the coordinates of x with respect to the canonical basis; for p infinite and $x \in V_p = \mathbb{R}^n$ one takes $\|x\|_p$ = square root of the sum of squares of the coordinates of x. Setting $G = GL(n)$ one denotes by K the compact subgroup of $G_A = GL(n, A)$ defined as follows: one sets $K = \prod K_p$ where $K_\infty = O(n)$ is the orthogonal group and where for finite p, $K_p = GL(n, \mathbb{Z}_p)$. Finally one denotes by $(e_i)_{1 \leq i \leq n}$ the canonical basis of k^n.

It is well-known (and elementary) that every matrix $g \in G_A = GL(n, A)$ can be written in the form $g = kt$ where $k \in K$ and where t is triangular. We will show that, *for every $g \in G_A$ there exists $\gamma \in G_k = GL(n, k)$ such that $g\gamma = kt$ where $k \in K$ and where t is a triangular matrix whose diagonal terms $t_i \in A^*$ satisfy the inequalities*

$$|t_i / t_{i+1}| \leq c = 2/\sqrt{3}.$$

In fact, by property (iv) of §1.1 one can, modulo G_k, suppose that $\|g(e_1)\| \leq \|g(\xi)\|$ for every non-zero $\xi \in V_k$; setting $g = kt$ ($k \in K$, t triangular such that $\|g(e_1)\| = |t_1|$, since the canonical height is invariant under K), and arguing by induction on n using the subgroup of G that fixes e_1 and stabilizes the subspace generated by the other basis vectors, one can suppose that the conditions $|t_i / t_{i+1}| \leq c$ are already realized for $i \geq 2$. Expressing that $\|g(e_1)\| \leq \|g(\xi)\|$ if ξ is a linear combination of e_1 and e_2 one gets

$$|t_1| \leq \|(\lambda + \mu u) t_1 e_1 + \mu t_2 e_2\|$$

for all $\lambda, \mu \in k$ not both zero, u being a certain coefficient of the matrix t; setting $x = t_1 / t_2$ yields $|x| \leq \|(u + v) x e_1 + e_2\|$ for every $v \in k$; since one does not change the two members of this inequality by multiplying x by an idele whose components all have absolute value 1, one sees that one can reduce to the case where x is the product of a non-zero element of k and an element of $k_\infty = \mathbb{R}$, then (by modifying u and v) to the case where $x_p = 1$ for every finite p; this done, one observes that, for every $u \in A$, there exists $v \in k$ such that $u_p + v_p$ is integer for every finite p and that $|u_\infty + v_\infty| \leq \frac{1}{2}$ (miracle for which the editor does not feel responsible); that yields then, for this value of v, the inequality $|x| \leq \sqrt{1 + |x|^2/4}$ which shows that $|t_1 / t_2| \leq c$ as claimed.

[It would be instructive to extend if possible the preceeding method to characteristic p, i.e. by taking as base field $k = \mathbb{F}_p(X)$ instead of the field of rationals.]

2 Mahler criterion

2.1 A property of properness

Let G be a linear algebraic group defined over k and H a closed subgroup of G defined over k. One obviously has a continuous injection of H_A / H_k into G_A / G_k; we will show that it is proper if restricted to H_A° / H_k, in other words (this is the same thing since this is a matter of locally compact spaces countable at infinity) that this application induces a *homeomorphism of H_A° / H_k onto a closed subspace of G_A / G_k*. We will give the proof for G connected, the general case immediately reducing to that.

In this case there exists in fact a vector space V defined over k, a representation ρ of G in V defined over k and a non-zero $a \in V_k$ such that H is the subgroup stabilizing the straight line generated by a. For every $h \in H_A$ one then obviously has $\rho(h)a = ta$ where $t \in A^*$, conversely, if $h \in G_A$ has this property, every component h_p of h leaves the straight line generated by a invariant, hence belongs to H_p and $h \in H_A$.

Let λ be the character of H given by $\rho(h)a = \lambda(h)a$ and H_A^λ the subgroup $|\lambda(h)| = 1$ of H_A; we will show that the image of H_A^λ / H_k in G_A / G_k is closed, otherwise said that $H_A^\lambda G_k$ is closed in G_A. Let $I^\circ(k)$ be the group of ideles of absolute value 1 of k; by the above $H_A^\lambda G_k$ is the inverse image of $I^\circ(k)\cdot\rho(G_k)a$ by the continuous application $g \to \rho(g)a$ of G_A into V_A; it thus suffices to show that $I^\circ(k)\cdot\rho(G_k)a$ is closed in V_A; but $I^\circ(k)$ is the product of a compact set with k^*, and $k^*\cdot\rho(G_k)a \subset V_k$ is closed since discrete.

One thus has a continuous injection of H_A^λ/H_k onto a closed subspace of G_A/G_k, as this is a matter of locally compact spaces countable at infinity this injection is a homeomorphism. As H_A°/H_k is obviously closed in H_A^λ/H_k the proof is complete.

2.2 The criterion of Mahler

Here it is:

THEOREM 1. *Let V be a vector space defined over k and G a closed subgroup of $GL(V)$ defined over k. In order that a subset M of G_A° is relatively compact modulo G_k it is necessary and sufficient that the following property is satisfied: given elements $m_n \in M$ and $\xi_n \in V_k$ such that $m_n(\xi_n)$ goes to 0 in V_A, one has $\xi_n = 0$ for n large.*

First the result of §2.1 allows one immediately to restrict to the case where $G = GL(V)$; we suppose in the following that V is the canonical vector space of dimension n, thus that $G = GL(n)$.

Suppose that M is compact modulo G_k; if $m_n(\xi_n)$ goes to 0 one can write $m = k_n\gamma_n$ where the k_n stay in a compact subset of G_A and where $\gamma_n \in G_k$; then it is clear that $\gamma_n(\xi_n) \in V_k$ goes to 0, hence is zero for large n by discreteness...

Suppose conversely that the condition is satisfied. Property (iii) of §1.1 shows immediately that there exists $c > 0$ such that $\|m(\xi)\| \geq c$ for every $m \in M$ and every non-zero $\xi \in k^n$; as we argue modulo G_k we can suppose $m = kt$ with $k \in K$ and t triangular such that $|t_i/t_{i+1}| \leq c'$ (cf. §1.2), one has $|t_1| = \|m(e_1)\| \geq c$ and on the other hand $|t_1\cdots t_n| = 1$ as $M \subset G_A^\circ$; thus the bounds

$$0 < c_1 \leq |t_i| \leq c_2 < +\infty$$

so that the t_i stay in a compact subset of A^* modulo k^*, and since the triangular matrices satisfying these conditions obviously form a compact subset modulo the rational triangular matrices, the proof is finished.

3 Where one gets rid of solvable groups

THEOREM 2 (ONO). *Let G be a solvable linear algebraic group defined over k; the space G_A°/G_k is compact.*

Let ρ be a linear representation of G in a vector space V, all defined over k; everything amounts to show that if elements $g_n \in G_A^\circ$ and $\xi_n \in V_k$ are such that $\rho(g_n)\xi_n$ converges to 0 in V_A, then $\xi_n = 0$ for large n (in fact, once this is established, it remains to apply the Mahler criterion by taking ρ faithful). We will argue by induction on $\dim(V)$, the case where $\dim(V) = 0$ not being difficult.

Let us consider a polynomial P on V that is homogeneous and semi-invariant under G, i.e. such that

$$P[\rho(g)x] = \chi(g)P(x)$$

where χ is a rational character of G and let us suppose that P is defined over k (in which case χ is also). The map from V_A to A defined by P is continuous, hence $P[\rho(g_n)\xi_n]$ converges to 0 in A; in view of (ii) in §1.1 the same then holds for

$$|P[\rho(g_n)\xi_n]| = |\chi(g_n)| \cdot |P(\xi_n)| = |P(\xi_n)|;$$

thus $P(\xi_n)$ is rational and therefore zero or of absolute value 1; consequently one has $P(\xi_n) = 0$ for every n sufficiently large.

To finish the proof, note that since G is solvable one can find a semi-invariant F homogeneous and of degree 1 defined on the algebraic closure $\overline{k}$ of k (take an eigenvector of G in the dual of V). We apply the result already obtained to the polynomial $P = \prod F^\sigma$ where σ describes the Galois group of $\overline{k}$ over k modulo the subgroup fixing F; one then sees that for n sufficiently large, ξ_n belongs to the subspace $W = \cap \mathrm{Ker}(F^\sigma)$ of V and the induction hypothesis shows that $\xi_n = 0$ for n sufficiently large, which finishes the proof.

4 A remark on the passage to the adjoint group

THEOREM 3. *Let G be a linear algebraic group defined over k and G^* its adjoint group. Then the canonical map $G^\circ_A/G_k \to G^*_A/G^*_k$ is proper.*

Suppose first that G is *reductive and connected.* G being realized as a closed subgroup of $GL(V)$ where V is a vector space defined over k, we denote by M the closed associative algebra formed by linear combinations of elements of G and by Z the multiplicative group of invertible elements of the center of M. Obviously Z is a linear algebraic group defined over k which commutes with G, and one can thus form $H = G \cdot Z$, a closed subgroup of $GL(V)$ defined over k. It is clear that $G \cap Z$ is the center of G, i.e. the kernel of the adjoint representation as G is connected, and consequently one has canonical isomorphisms $G^* = G/G \cap Z = H/Z$. In order to show that the map $G^\circ_A/G_k \to G^*_A/G^*_k$ is proper it thus suffices, as the composition of two proper maps is proper, to show that this also holds for the maps $G^\circ_A/G_k \to H^\circ_A/H_k$ and $H^\circ_A/H_k \to G^*_A/G^*_k$. For the first, this results from §2.1. For the second, we note that G being reductive its enveloping algebra M is semi-simple, hence also the center of M. Theorem 90 thus shows that if k' is a Galois extension of k with Galois group Σ one has $H^1(\Sigma, Z_{k'}) = 0$. As the rational points are dense in H/Z (we are in characteristic 0) we deduce from this, by a known argument [6], that the fibration of H by Z is locally trivial and consequently that H°_A/H_k is a principal fibre bundle with base G^*_A/G^*_k and structural group Z°_A/Z_k that is *compact* in view of Theorem 2, hence the properness. [N.B. This argument uses the fact that $Z^\circ_A = Z_A \cap H^\circ_A$, which holds since the restrictions of rational characters of H defined over k to Z form a finite index subgroup of the group of rational characters of Z defined over k, a result which holds for every reductive group.]

The theorem being established for reductive connected G, it remains to examine the general case. For that, one reduces to the reductive case by passing to the quotient by the unipotent radical (a harmless operation as the fibration by a unipotent subgroup is always locally trivial), and the reductive case reduces to the connected reductive case by observing that if H is the connected component of G, the quotient G_A/H_A is compact (but not finite...).

5 Anisotropic reductive groups

A reductive group G defined over k is called *anisotropic* if every $g \in G_k$ is semi-simple, i.e. if no element of G_k is unipotent (the decomposition in semi-simple and unipotent holds over k); it amounts to the same to require that every rational element over k of the derived algebra of the Lie algebra of G is semi-simple, i.e. that this derived algebra does not contain any rational element over k and nilpotent. Thus to say that G is anisotropic amounts to say that the adjoint group is anisotropic.

THEOREM 4 (BOREL-HARISH-CHANDRA). *Let G be a connected reductive group defined over k. In order that G_A°/G_k is compact it is necessary and sufficient that G is anisotropic.*

In view of Theorem 3 one can pass to the adjoint group, i.e. assume that G is the group of automorphisms of a semi-simple Lie algebra $\mathfrak{g}$ defined over k, in which case we will prove that G_A/G_k is compact. By Theorem 1 it suffices to show that if one has elements $g_n \in G_A$ and $\xi_n \in \mathfrak{g}_k$ such that $g_n(\xi_n)$ goes to 0 in $\mathfrak{g}_A$, then $\xi_n = 0$ for n large. Now let us denote by P an arbitrary coefficient of the characteristic polynomial of $\mathrm{Ad}(X)$, where X is a generic element of $\mathfrak{g}$. Then P is a polynomial on $\mathfrak{g}$ with coefficients in k and invariant under G; and $P(g_n(\xi_n)) = P(\xi_n)$ goes to $P(0)$ so that $P(\xi_n) = P(0)$ for n large. In other words, the operator $\mathrm{Ad}(\xi_n)$ is nilpotent for large n and as $\mathfrak{g}_k$ does not contain non-zero nilpotent elements, one concludes that $\xi_n = 0$, which finishes a beautiful proof due to MOSTOW and TAMAGAWA [4].

Concerning the fact that G is anisotropic if G_A/G_k is compact, we first observe that the second property implies that, for every $\xi \in \mathfrak{g}_k$, the orbit $G_A(\xi)$ is closed in G_A; but if G were not anisotropic one could take ξ nilpotent and then there would exist a subgroup of G (closed and defined over k) stabilizing the straight line generated by ξ without stabilizing the vector ξ itself (cf. [2], p. 21); hence $G_A(\xi)$ would contain the "straight line" $A^* \cdot \xi$ so that only 0 would be adherent to $G_A(\xi)$, which is absurd.

[The argument of Theorem 4 applies directly, i.e. without passage to the adjoint group, if G is embedded in a $GL(V)$ in such a way that the orbit $G(\xi)$ of every $\xi \in V_k$ is closed in the sense of algebraic geometry. In fact, this hypothesis immediately allows to "separate" 0 from the $\xi \neq 0$ with the help of the invariants of G with coefficients in k, and one then argues as MOSTOW-TAMAGAWA. It would be interesting to know if every anisotropic group admits a *faithful* representation of this type, including characteristic p. In general, it seems that one does not know much on these questions of orbits, which nevertheless should be interesting from all point of views.]

6 General reductive groups: simple roots

The goal of this paragraph is to supply (without proofs) certain indispensable complements to a previous expose [3]. See [1].

A. Let G be a connected reductive group defined over k and not anisotropic so that G_k contains unipotent elements different from e. By [3], Theorem 9, there exist tori in G defined and split over k and not contained in the center of G; we denote by T a torus defined and split over k and maximal with respect to this conditions; T is unique modulo conjugation by an element of G_k ([3], Corollary 1 of Theorem 6). Let $\mathfrak{g}$ be the Lie algebra of G (we excuse for this recourse to characteristic 0...) and, for every character α of T, let $\mathfrak{g}(\alpha)$ be the subspace of those $X \in \mathfrak{g}$ for which $\mathrm{Ad}(t)X = \alpha(t)X$ for all $t \in T$; then every $\mathfrak{g}(\alpha)$ is defined over k and $\mathfrak{g}$ is the direct sum of the non-zero $\mathfrak{g}(\alpha)$. Those α for which $\mathfrak{g}(\alpha)$ is not zero are the *roots* of G with respect to T (we include the unit character among the roots), these are obviously the restrictions to T of the roots of G with respect to a maximal torus containing T.

Let $r = \dim(T/T \cap Z)$ where Z is the center of G — we will say that r is the *rank* of G over k. Then there exist r linearly independent roots $\alpha_1, \dots, \alpha_r$ such that every root of G with respect to T is a linear combination of the α_i with integer coefficients all of the same sign; once such a system of simple roots is chosen, one can talk of "positive" roots (and conversely, in order to construct a system of simple roots one totally orders the roots and one considers the non-decomposable positive roots). We will denote by U the subgroup of G whose Lie algebra is the sum of the $\mathfrak{g}(\alpha)$ with $\alpha > 0$, by $Z(T)$ — this obviously is the centralizer of T — the subgroup corresponding to $\mathfrak{g}(0)$, and by P the subgroup generated by $Z(T)$ and U; one easily shows that P is a *minimal parabolic subgroup* of G in the sense of [3], that U is its unipotent radical and is a maximal nilpotent subgroup defined over k of G, finally that P is the semidirect product of U and $Z(T)$. The group $Z(T)$ is reductive, has center T, and is anisotropic because T, which is a maximal split subtorus of it, is in its center.

B. For every i such that $1 \leq i \leq r$, we consider the roots $\alpha = \sum n_j(\alpha)\alpha_j$ for which one has $n_i(\alpha) \geq 0$; the sum of the $\mathfrak{g}(\alpha)$ corresponding to these roots is the Lie algebra of a subgroup $P(i)$ of G, defined over k and containing P. One has $P(i) = Z(i)U(i)$ (a semidirect product over k) where $Z(i)$ is the reductive subgroup generated by the roots for which $n_i(\alpha) = 0$, and $U(i) \subset U$ the unipotent group generated by the roots for which $n_i(\alpha) \geq 1$. It is clear that $Z(i)$ is of rank $r - i$; it admits T as a maximal split torus, $P \cap Z(i)$ as a minimal parabolic subgroup, and the simple roots of $Z(i)$ with respect to T and $Z(i) \cap P$ are the α_j $(j \neq i)$. The role played by the $Z(j)$ in G is played in $Z(i)$ by the $Z(i) \cap Z(j)$.

For every i we will denote by $V(i)$ a vector space defined over k, by ρ_i a representation of G defined over k in $V(i)$, and by a_i a non-zero rational vector in $V(i)$, chosen in such a way that $P(i)$ is the stabilizer of the straight line generated by a_i; one then has $\rho_i(p)a_i = \Delta_i(p)a_i$ where Δ_i is a character of $P(i)$ defined over k ("dominant weight" of ρ_i); we will say that the ρ_i form a *system of fundamental representations of G with respect to P* (a non-orthodox terminology: the fundamental representations are usually assumed to have dominant weights Δ_i as low as possible, a condition we do not impose here). It is clear that, for every i, the restrictions of the ρ_j $(j \neq i)$ to $Z(i)$ form a system of fundamental representations of $Z(i)$ with respect to $Z(i) \cap P$.

If G has rank 1 over k, there is one fundamental representation ρ_1, and there are two *strictly positive* integers a and b as well as a character χ of G defined over k such that one has on T the relation

$$\Delta_1(t)^a = \chi(t) \cdot \alpha_1(t)^b;$$

one then deduces that

$$|\Delta_1(t)| = |\alpha_1(t)|^s \quad \text{for every} \quad t \in T_A \cap G_A^\circ$$

where s is a *strictly positive* rational number.

Finally, the "theorem of Bruhat" has been extended by BOREL and TITS to the case studied here in the following way: let $N(T)$ be the normalizer of T in G and $W(T) = N(T)/Z(T)$, obviously a finite group; then every element of $W(T)$ is represented by an element of $N(T)_k$ and G_k is the union of double cosets $U_k v P_k$ where v denotes $N(T)_k$ modulo $Z(T)_k$. We will not need this result here.

7 Finiteness of the class number

The results of §5 and §6 already allow to establish in full generality the theorem of "finiteness of the class number". Let G be a closed subgroup of $GL(V)$ defined over k, where V is a vector space defined over k. Let us choose a basis of V_k over k and for every finite p let K_p be the open compact subgroup of G_p which fixes the p-adic lattice generated by this basis. Denoting by α the subgroup of V_k generated by this basis, one then has in the adelic group G_A an *open* subgroup

$$W(\alpha) = G_\infty \times \prod_p K_p.$$

The result we have in mind in this paragraph is the following:

THEOREM 5. *The set $W(\alpha)\backslash G_A/G_k$ is finite.*

This is obvious if G_A/G_k is compact (BOREL-LEBESGUE...), by passing to the quotient by the unipotent radical one immediately reduces the claim to the case where G is reductive. Using §6 one first sees that G_A/P_A is *compact* because the fibration of G by P is locally trivial ([3], Theorem 7), thus $G_A/P_A \simeq (G/P)_A$ is the variety of adeles of a complete variety and even projective, thus compact. Since $W(\alpha)$ is open, there are finitely many elements x_i of G_A such that

$$G_A = \bigcup W(\alpha) x_i P_A.$$

Denote by α_i the lattice obtained from α by transforming it with x_i, as $x_i W(\alpha_i) = W(\alpha)x_i$ one is obviously led to show that P_A is a finite union of double cosets modulo $W(\alpha_i) \cap P_A$ from the left and P_k from the right. But $W(\alpha_i) \cap P_A$ plays the same role for P_A as $W(\alpha)$ for G_A, thus this reduces to establishing Theorem 5 for P, i.e. for P modulo its unipotent radical, i.e. for a reductive *anisotropic* group and then the claimed result is a consequence of Theorem 4 and the obvious fact that

$$G_A = G_\infty G_A^\circ.$$

Note that we in fact proved that G_A is the finite union of classes $W(\alpha)xP_k$. We will deduce the following result:

THEOREM 6. *Let V be a vector space defined over k, α a lattice in V_k, G a connected closed subgroup of $GL(V)$ defined over k, and P a minimal parabolic subgroup of G. Let Γ be the subgroup of G_k which fixes the lattice α. Then $\Gamma\backslash G_k/P_k$ is finite.*

One obviously has $\Gamma = G_k \cap W(\alpha)$. Let us write $G_A = \cup W(\alpha)x_i P_k$ with a finite number of x_i, and in particular $G_k \subset \cup W(\alpha)x_i P_k$; it is clear that G_k is contained in the union of those classes $W(\alpha)x_i P_k$ which actually meet G_k, and for such a class one can suppose that $x_i = \xi_i \in G_k$; this trivially yields $G_k = \cup \Gamma \xi_k P_k$ and thus the theorem.

From Theorem 6 one immediately deduces the following results:

(i) *there are finitely many minimal parabolic subgroups of* G *modulo conjugation by elements of* Γ;

(ii) *if* U *is a maximal unipotent subgroup of* G *defined over* k, *then there are finitely many subgroups* $\Gamma \cap \xi U \xi^{-1}$ $(\xi \in G_k)$ *of* Γ *modulo conjugation by elements of* Γ (*we note that the* $\xi U \xi^{-1}$ *are the different maximal unipotent subgroup of* G *defined over* k).

8 Minkowski inequalities (stated)

From now on G is a connected reductive group of rank $r \geq 1$ over k and we use the notations of §6. Our goal is to construct open sets Ω in G_A defined by inequalities analogous to those of §1.2 and such that $G_A = \Omega \cdot G_k$, which in particular shows that G_A° / G_k *is of finite volume.*

For that end we choose once for all a compact set M in G_A such that $G_A = M \cdot P_A$ and an open relatively compact set F in P_A° such that $P_A^\circ = F \cdot P_k$ (this is possible in view of the fact that $P_A^\circ = Z(T)_A^\circ U_A$ and that $Z(T)$ is anisotropic); finally, for every $c > 0$, denote by $T_A(c)$ the part of T_A defined by the inequalities $|\alpha_i(t)| < c$; we then set

$$\Omega(c) = M \cdot T_A(c) \cdot F.$$

The first result we have in mind is the following:

THEOREM 7 (BOREL). *There exist a* $c > 0$ *such that one has* $G_A = \Omega(c) \cdot G_k$.

(This result immediately implies that G_A° / G_k is of finite volume; in fact take the K_p of §7 and let K_∞ be a maximal compact of G such that one has $G_\infty = K_\infty P_\infty$ as we will soon see; Theorem 5 then shows that we can take for M a finite union of classes Kx where K is the product of the K_p for p finite or not; one has thus to show $KxT_A(c)F \cap G_A^\circ$ has finite measure; but on the open class KxP_A the Haar measure of G_A is, up to a factor, the product of that of K by the right invariant measure of P_A, and the latter is the product of the measures invariant under $Z(T)_A$, U_A and a factor which is the determinant of the (adelic!) adjoint representation of $Z(T)_A$ in the Lie algebra of U_A, this determinant is a monome with strictly *positive* integer exponents in the simple roots α_i; finiteness of volume follows at once.)

The "reduction" indicated by Theorem 7 can otherwise be obtained by a method analogous to the one of §1.2. We consider the representations ρ_i of §6 and choose on every $V(i)$ a height (§1.1). For every $c > 0$ denote by $\mathscr{R}(c)$ the set of elements $g \in G_A$ which verify the following conditions:

$$\|\rho_1(g)a_1\| \leq c\|\rho_1(g\gamma)a_1\| \quad \text{for every} \quad \gamma \in G_k,$$
$$\|\rho_2(g)a_2\| \leq c\|\rho_2(g\gamma)a_2\| \quad \text{for every} \quad \gamma \in P(1)_k,$$
$$\|\rho_3(g)a_3\| \leq c\|\rho_3(g\gamma)a_3\| \quad \text{for every} \quad \gamma \in P(1)_k \cap P(2)_k$$

and so on (the γ's allowed in every inequality are of course those which do not destroy the previous inequalities...). By §1.1 (iv) one *obviously* has $G_A = \mathscr{R}(1) \cdot G_k$; this said, we will establish also the following result (which implies Theorem 7 trivially, as we saw - but one could not directly prove Theorem 8 without going, at least in parts, through a proof of Theorem 7):

THEOREM 8. *For every $c > 0$ there exist a $c' > 0$ such that one has $\mathscr{R}(c) \subset \Omega(c')P_k$ (it is evident that $\mathscr{R}(c)$ is right invariant under P_k).*

9 Proofs of Theorems 7 and 8

9.1 Proof of Theorem 7 (rank 1)

In this section we suppose that G has rank 1. Taking into account that

$$P_A = T_A P_A^\circ, \quad T_A \cap P_A^\circ = T_A^\circ$$

(reduce this to $Z(T)$ and observe that the characters of $Z(T)$ defined over k are determined, up to a finite number of choices, by their restrictions to the maximal split torus T), one sees that there exists a strictly positive, continuous function $d(g)$ on G_A that satisfies

$$d(gtp) = d(g)|\alpha_1(t)| \quad \text{for} \quad t \in T_A \quad \text{and} \quad p \in P_A^\circ.$$

Considering the compact set M such that $G_A = M \cdot P_A$ chosen in order to define the $\Omega(c)$, it is clear that $d(M)$ is a compact subset of $\mathbb{R}_+^*$ and hence that, for $g = mtp$ ($m \in M, t \in T_A, p \in P_A^\circ = F \cdot P_k$) the numbers $d(g)$ and $|\alpha_1(t)|$ are "equivalent". Consequently, Theorem 7 is equivalent to the relation

$$\sup_{g \in G_A} \inf_{\gamma \in G_k} d(g\gamma) < +\infty;$$

it obviously suffices by the way to vary g above in the set E of those g such that $d(g\gamma) \geq 1$ for every $\gamma \in G_k$, and the claim will not less obviously result from the fact that $E \cap G_A^\circ = E^\circ$ *is compact* modulo G_k.

It remains to establish this last point, and for that one can replace E° by its image in the adjoint group (Theorem 3); thus (Theorem 1) it remains to show that if elements $g_n \in E^\circ$ and rational elements ξ_n in the Lie algebra $\mathfrak{g}$ of G with $\mathrm{Ad}(g_n)\xi_n$ going to 0 in $\mathfrak{g}_A$, then $\xi_n = 0$ for large n. One first sees, by reasoning as in the proof of Theorem 4, that one has $P(\xi_n) = P(0)$ for every invariant P of the adjoint representation, in particular for the coefficients of the characteristic polynomial of Ad(?); thus the $\mathrm{Ad}(\xi_n)$ are nilpotent for large n: writing $\mathfrak{g}$ as direct sum of its center and the derived algebra $\mathfrak{g}'$, and observing that the central component of ξ_n is zero (as $\mathrm{Ad}(g_n)$ acts trivially on the center of $\mathfrak{g}$) one sees that in addition $\xi_n \in \mathfrak{g}_k'$ for large n; thus the ξ_n generate in G unipotent one-parameter subgroups defined over k, thus can be transformed in subgroups of U by G_k, and thus, by replacing g_n by $g_n\gamma_n$ with $\gamma_n \in G_k$ if necessary, one can assume that the ξ_n are in the Lie algebra of U, i.e. write

$$\xi_n = \xi_n' + \xi_n'' \quad \text{with} \quad \xi_n' \in \mathfrak{g}(\alpha_1) \quad \text{and} \quad \xi_n'' \in \mathfrak{g}(2\alpha_1).$$

Let us write now $g_n = k_n t_n z_n u_n$ where the k_n stay in a fixed compact subset, and where $t_n \in T_A, z_n \in Z(T)_A^\circ, u_n \in U_A$; as the k_n stay in a fixed compact subset, the $\mathrm{Ad}(t_n z_n u_n)\xi_n$ go to 0; but in view of the relation $[\mathfrak{g}(\lambda), \mathfrak{g}(\mu)] \subset \mathfrak{g}(\lambda + \mu)$, it is clear, that one has

$$\mathrm{Ad}(t_n z_n u_n)\xi_n \equiv \alpha_1(t_n)\mathrm{Ad}(z_n)\xi'_n \bmod \mathfrak{g}(2\alpha_1),$$

so that $\alpha_1(t_n)\mathrm{Ad}(z_n)\xi'_n$ goes to 0; as by assumption $\alpha_1(t_n)$ remains larger than 1, one sees that $\mathrm{Ad}(z_n)\xi'_n$ goes to 0, but as $Z(T)$ is anisotropic Theorem 4 shows that one can suppose that the z_n stay in a fixed compact subset — and then ξ'_n goes to 0 and thus is zero for large n. This done, it remains to kill the second component ξ''_n, which one does by observing that

$$\mathrm{Ad}(t_n z_n u_n)\xi''_n = \alpha_1(t_n)^2\mathrm{Ad}(z_n)\xi''_n$$

and by reasoning as above.

9.2 Proof of Theorem 8 (rank 1)

We always assume that G has rank 1 and we set $d'(g) = \|\rho_1(g)a_1\|$. As we saw in §6 there exist positive integers r and s and a character χ of G defined over k, such that

$$\Delta_1(t)^r = \chi(t)\cdot\alpha_1(t)^s;$$

the function $d'(g)^r/|\chi(g)\cdot|d(g)^s$ is thus right invariant under P_A, and as G_A/P_A is compact one deduces that there exist constants $c', c'' > 0$ such that one has

$$c'|\chi(g)|\cdot d(g)^s \le d'(g)^r \le c''|\chi(g)|\cdot d(g)^s$$

for every $g \in G_A$. Let us thus suppose that $g \in \mathscr{R}(c)$, i.e. $d'(g) \le c\cdot d'(g\gamma)$ for every γ; we then obviously have

$$c'|\chi(g)|d(g)^s \le \inf_\gamma |\chi(g\gamma)|d(g\gamma)^s;$$

but one has $|\chi(g\gamma)| = |\chi(g)\chi(\gamma)| = |\chi(g)|$ in view of the product formula, and on the other hand the already proved Theorem 7 shows that $\inf d(g(\gamma))$ is below a fixed constant; one immediately deduces that the function $d(g)$ has an upper bound on $\mathscr{R}(c)$, which obviously proves Theorem 8.

9.3 Proof of Theorems 7 and 8 (general case)

As Theorem 7 obviously is a consequence of Theorem 8, we will restrict ourselves to prove the latter by induction on the rank r of G.

We consider the compact set M of §8, such that $G_A = M\cdot P_A$, and for every $g \in \mathscr{R}(c)$ we write $g = mp$ with $m \in M$ and $p \in P_A$; by property (v) in §1.1 the ratios $\|\rho_i(g\gamma)a_i\|/\|\rho_i(p\gamma)a_i\|$ stay in a fixed compact subset of $\mathbb{R}_+^*$ and consequently there is c' such that $p \in \mathscr{R}(c')$. Setting $p = tp'$ with $t \in T_A$ and $p' \in P_A^\circ = F\cdot P_k$ one sees that it remains to prove the existence of a c'' such that $|\alpha_i(t)| < c''$ for every i.

We consider the reductive subgroup $Z(1)$ of rank $r-1$ defined in §6 and set $p = zu$ with $z \in Z(1)_A$ and $u \in U(1)_A$; the hypothesis $p \in \mathscr{R}(c)$ implies, as $U(1)$ is invariant under $Z(1)$ and fixes the a_i, the inequalities

$$\|\rho_2(z)a_2\| \le c\|\rho_2(z\zeta)a_2\| \quad \text{for every} \quad \zeta \in Z(1)_k,$$

$$\|\rho_3(z)a_3\| \le c\|\rho_3(z\zeta)a_3\| \quad \text{for every} \quad \zeta \in Z(1)_k \cap P(2)_k$$

and so on, and hence z remains in a $\mathscr{R}(c)$ of the subgroup $Z(1)$; the induction hypothesis then shows that the $|\alpha_i(t)|$ remain bounded from above for $i \geq 2$.

It remains to bound $|\alpha_1(t)|$. To that end we use the subgroup H (reductive of rank 1) of G generated by the roots proportional to α_1 and the unipotent subgroup V generated by the $\alpha > 0$ not proportional to α_1; it is clear that H normalizes V, contains $Z(T)$ and that P is the semidirect product of V and the minimal parabolic subgroup $H \cap P$ of H; finally ρ_1 obviously is a fundamental representation of H. This said, if one writes $p = thv$ with $v \in V_A$ and $h \in (H \cap P)_A^\circ$ the hypothesis that $p \in \mathscr{R}(c)$ implies an estimate of the form

$$\|\rho_1(th)a_1\| \leq c\|\rho_1(th\gamma)a_1\| \quad \text{for every} \quad \gamma \in H_k;$$

as H has rank 1, one concludes that $|\alpha_1(t)| \leq c'$, and this finishes the proof.

10 Construction of open fundamental sets

Theorem 7 shows that, for c sufficiently large, G_A is the union of open sets $\Omega(c)\gamma$ where γ runs through G_k; our goal now is to construct in G_A *open fundamental sets* for G_k, i.e. open sets σ such that one has

$$G_A = \sigma \cdot G_k$$

and such that moreover the number of those $\gamma \in G_k$ for which the intersection $\sigma\gamma \cap \sigma$ is not empty is finite, a property which in general does not hold for the $\Omega(c)$ of §8.

To that end we call *open Siegel set* in G_A every open set of the form

$$\sigma = M \cdot T_\infty^+(c) \cdot F$$

constructed as follows: $T_\infty^+(c)$ is the set of $t \in T_\infty^+$ (connected component of the identity in the Lie group T_∞) such that one has $\alpha_i(t) < c$ for $1 \leq i \leq r$; F is an open relatively compact subset of P_A° such that one has $P_A^\circ = F \cdot P_k$; finally M is a compact set such that $G_A = M \cdot P_A$ and chosen in such way that the image M_∞ of M under the canonical map from G_A to G_∞ is a *maximal compact subgroup of* G_∞ *adapted to* T_∞ (i.e. such that the Lie algebras of M_∞ and T_∞ are orthogonal with respect to the Killing form of $\mathfrak{g}$ and also such that the symmetry with respect to M_∞ restricts to $t \to t^{-1}$ on T_∞). The existence of such an M is quite evident if one takes into account on the one hand that G_A/P_A is compact and on the other hand the fact that for *every* maximal compact subgroup M_∞ of G_∞ holds $G_\infty = M_\infty P_\infty$ by the classical Iwasawa theorem.

The assumption made on M_∞ comes into play only in the last lines of the proof (and in particular does not play any role in the Lemmas 1 to 4 below).

If one compares the open set σ with the previous open set $\Omega(c) = M \cdot T_A(c) \cdot F$ one first observes that one has

$$T_A(c) = T_\infty^+(c)T_A^\circ \subset T_\infty^+(c)P_A^\circ = T_\infty^+(c)FP_k$$

which yields $\Omega(c) \subset \sigma \cdot P_k$; Theorem 7 thus shows that

$$G_A = \sigma G_k$$

as soon as c is sufficiently big. In order to exhibit open fundamental sets it remains to establish the following result:

THEOREM 9 (BOREL). *Let σ' and σ'' be two open Siegel sets in G_A; then the set of $\gamma \in G_k$ such that $\sigma'\gamma$ meets σ'' is finite.*

The proof consists in examining first all triples (g', g'', γ) such that one has

$$g' \in \sigma', \quad g'' \in \sigma'', \quad \gamma \in G_k \quad \text{and} \quad g'\gamma = g''$$

and for which γ belongs to a given double coset class modulo P_k, in other words (theorem of Bruhat-Borel-Tits) for which

$$\gamma = \pi' w \pi''$$

with $w \in N(T)_k$ given, $\pi', \pi'' \in P_k$ varying.

In what follows we set

$$g' = m't'p', \quad g'' = m''t''p''$$

with $m', m'' \in M$, $t', t'' \in T^+_\infty(c)$ and $p', p'' \in F$.

LEMMA 1. *For every $c > 0$ and every compact subset F of P_A the set of elements tpt^{-1} $(t \in T^+_\infty(c), p \in F)$ is relatively compact in P_A.*

Since p and tpt^{-1} are obviously equal at every finite place it suffices to consider only G_∞. But $p = zu$ where z stays in a compact subset of $Z(T)_\infty$ and u in a compact subset of U_∞, and as $tpt^{-1} = ztut^{-1}$ it suffices to examine tut^{-1}. Setting $u = \exp(X)$, where X belongs to the direct sum of the $\mathfrak{g}(\alpha)_\infty$ for all roots $\alpha > 0$, it obviously suffices to show that if X stays in a fixed compact subset then this also holds for $\mathrm{Ad}(t)X$. To that end we can reduce to the case where $X \in g(\alpha)_\infty$, and thus $\mathrm{Ad}(t)X = \alpha(t)X$; but $\alpha(t)$ is a monome in the $\alpha_i(t)$ with *positive* integer exponents, thus stays bounded on $T^+(c)$, which implies the lemma.

LEMMA 2. *The element $w^{-1}t'wt''^{-1}$ stays in a fixed compact subset of T^+_∞.*

The relation $g'\gamma = g''$ can actually be written as

$$m't'p't'^{-1} \cdot t'\gamma = m''t''p''t''^{-1} \cdot t''$$

and shows, in view of Lemma 1, that

$$t'\gamma t''^{-1} = t'\pi' w \pi'' t''^{-1}$$

stays in a fixed compact set.

We consider now a vector space V defined over k, a linear representation ρ of G in V defined over k and a non-zero vector $a \in V_k$ such that one has

$$\rho(p)a = \Delta(p)a$$

where Δ is a character of P defined over k. If we choose a height on V_A, property (vi) of §1.1 shows that one has [1]

[1] The relation $x \asymp y$ means that the ratio x/y stays in compact subset of $\mathbb{R}^*_+$ and the relation $x \prec y$ means that it remains bounded from above.

$$\|\rho(t'\pi' w\pi'' t''^{-1})a\| \asymp 1\,,$$

as $|\Delta(\pi'')| = 1$, this can also be written in the form

$$\Delta(t''^{-1})\|\rho(t'\pi' w)a\| \asymp 1\,;$$

writing $t'\pi' w = t'\pi' t'^{-1} w w^{-1} t' w$ we thus obtain

$$\Delta(t''^{-1})\Delta(w^{-1}t'w)\|\rho(t'\pi' t'^{-1})\rho(w)a\| \asymp 1\,,$$

i.e.

$$\Delta(w^{-1}t'wt''^{-1})\|\rho(t'\pi' t'^{-1})\rho(w)a\| \asymp 1. \tag{1}$$

But $t'\pi' t'^{-1}$ stays in $P_A^\circ = F\cdot P_k$ with F relatively compact; applying again property (vi) of §1.1 and taking into account the fact that the heights of the non-zero rational elements of V stay away from 0 (property (iv) of §1.1) one sees that $\|\rho(t'\pi' t'^{-1})\rho(w)a\| \succ 1$ and consequently (1) leads to the relation

$$\Delta(w^{-1}t'wt''^{-1}) \prec 1. \tag{2}$$

We will show that one actually also has the opposite estimate

$$\Delta(w^{-1}t'wt''^{-1}) \succ 1\,, \tag{3}$$

in other words, in view of (1), that one also has

$$\|\rho(t'\pi' t'^{-1})\rho(w)a\| \prec 1\,. \tag{4}$$

In fact, first it is clear that the vector $\xi = \rho(w)a$ belongs to the weight $\Delta(w^{-1}t'w)$ so that (4) is equivalent to the relation

$$\Delta(w^{-1}t'w)^{-1}\|\rho(t'\pi')\xi\| \prec 1\,;$$

but as $\pi' \in P_k$ and as P is generated by the roots $\alpha \geq 0$, it is clear that one has a relation of the form

$$\rho(\pi')\xi = \sum \xi_\lambda$$

where

$$\rho(t')\xi_\lambda = \Delta(w^{-1}t'w)\lambda(t')\xi_\lambda$$

and where every λ is a monome in the $\alpha_i(t)$ with *positive* integer exponents; the relation (3) or (4) to establish can thus be written as

$$\|\sum \lambda(t')\xi_\lambda\| < 1\,,$$

which easily results from the fact that, as t' stays in $T^+(c)$, the $\lambda(t')$ remain bounded from above.

This done, by comparing (2) and (3) one sees that one has $\Delta(w^{-1}t'wt''^{-1}) \asymp 1$ whenever Δ is a "dominant weight" of G with respect to P; but one knows that the restriction to T of these dominant weights generate a subgroup of finite index of the group of all rational characters of T; this immediately yields

$$\chi(w^{-1}t'wt''^{-1}) \asymp 1$$

for every rational character χ of T, which obviously proves Lemma 2.

LEMMA 3. *For every j such that $1 \le j \le r$ either holds $\alpha_j(t') \succ 1$ or $w \in P(j)$.*

As the Weyl group permutes the roots one in fact has for every i a relation of the form

$$\alpha_i(w^{-1}t'w) = \prod \alpha_j(t')^{n_{ij}(w)}$$

with integers $n_{ij}(w)$ all positive or all negative. But one has the relations

$$\alpha_j(t') \prec 1, \quad \alpha_i(w^{-1}t'w) \prec 1$$

(the first one since t' stays in $T^+_\infty(c')$, the second one in view of Lemma 2 and the fact that t'' stays in $T^+(c'')$); it is thus clear that, if the $\alpha_j(t')$ do not stay away from zero, one has $n_{ij}(w) \ge 0$ for every i, which means that the transform of the root α_i is in $P(j)$ for every i, or moreover that $wUw^{-1} \subset P(j)$; but then wUw^{-1} is a maximal unipotent subgroup of $P(j)$ defined over k, thus conjugate to U by a rational element of $P(j)$ and consequently

$$w \in P(j)_k N(U)_k = P(j)_k P_k = P(j)_k$$

which finishes the proof.

LEMMA 4. *Suppose that w does not belong to any parabolic subgroup Q of G such that $P \subset Q$ and $Q \ne G$. Then the number of $\gamma \in P_k w P_k$ for which $\sigma'\gamma$ meets σ'' is finite.*

In fact Lemma 3 shows that then $\alpha_j(t') \succ 1$ for every j, hence that

$$\alpha_j(t') \asymp 1$$

for every j which means that $t'' = t'_1 z'$ where t'_1 stays in a compact set and where $z' \in T_\infty \cap Z_\infty$ (where Z is the center of G); Lemma 2 then shows that $t'' = t''_1 z'$ with t''_1 in a fixed compact set. The relation $g'\gamma = g''$ can then be written

$$m' t'_1 z' p' \gamma = m'' t''_1 z' p''$$

i.e. $m' t'_1 p' \gamma = m'' t''_1 p''$ since z' is in the center of G_A, and one sees that γ stays in a fixed compact set, thus the lemma.

Lemma 4 shows that among the $\gamma \in G_k$ for which $\sigma'\gamma$ meets σ'' those who do not belong to a parabolic subgroup Q containing P and different from G are finite in number. To finish the proof it thus remains to see that, for every parabolic subgroup Q containing P, the $\gamma \in Q_k$ such that $\sigma'\gamma$ meets σ'' and which do not belong to a parabolic subgroup containing P and strictly contained in Q are finite in number.

But Q is the semidirect product of a reductive group $H \supset Z(T)$ and its unipotent radical $V \subset U$. Denoting generally by q_H and q_V the components of a $q \in Q_A$ in the decomposition $Q_A = H_A V_A$, the relation $g'\gamma = g''$, i.e. $m't'\rho'\gamma = m''t''p''$ or also

$$m''^{-1} m' t' p' \gamma = t'' p''$$

implies $m''^{-1} m' \in Q_A \cap M^{-1} M$ and decomposes in the relations

$$(m''^{-1}m')_H t' p'_H \gamma_H = t'' p''_H, \tag{5}$$

$$\gamma_H^{-1} p_H'^{-1} t'^{-1} (m''^{-1}m')_V t' p' \gamma = p''; \tag{6}$$

to show that the number of γ's in question is finite, it thus suffices to show that on the one hand the number of possible components γ_H is finite and on the other hand that $t'^{-1}(m''^{-1}m')_V t'$ stays in a fixed compact set.

To establish the first point, we distinguish two cases. If $H = Z(T)$, i.e. if $Q = P$, Lemma 2 applies with $w = e$ and shows that $t'^{-1}t''$ stays in a fixed compact set; but since t' commutes with $(m''^{-1}m')_H$ as one supposes that $H = Z(T)$, relation (5) can be written as $(m''^{-1}m')_H p'_H \gamma_H = t'^{-1}t'' p''_H$, and shows that γ_H remains in a fixed compact set, hence the result. If now $H \neq Z(T)$, Lemma 4 applies to H (with respect to T and $H \cap P$), and to deduce that γ_H can only take finitely many values it suffices, as one is in a situation where γ_H does not belong to any parabolic subgroup of H containing $H \cap P$ and different from H, to show that $(m''^{-1}m')_H t' p_H$ and $t'' p''_H$ stay in a part of H_A which is the product of a compact set with the set of those $t \in T^+_\infty$ where the simple roots of H with respect to T and $P \cap H$ are bounded from below by a fixed number and by a compact subset of $(H \cap P)^\circ_A = H_A \cap P^\circ_A$; which is clear as the simple roots of H are obviously certain of the α_i. In this way one has established that γ_H can take on only finitely many values (without needing the assumption on M_∞ made at the beginning of this paragraph; but it will be used now).

In fact, it remains to verify that $t'^{-1}(m''^{-1}m')_V t''$ stays in a compact set, which will be done by showing much more, namely

$$t'^{-1}(m''^{-1}m')_V t' = (m''^{-1}m')_V.$$

As t' is equal to 1 at every finite place, it suffices to see that $(m''^{-1}_\infty m'_\infty)_V = 1$, in other words that

$$M_\infty^{-1} M_\infty \cap Q_\infty \subset H_\infty$$

but this results exactly from the hypothesis that M_∞ is a maximal compact subgroup of G_∞ adapted to T. In fact, the symmetry with respect to M_∞ transforms every $t \in T^+_\infty$ in t^{-1}, thus every root α in its opposite $-\alpha$, and consequently maps Q_∞ onto the "symmetric" subgroup Q'_∞ generated by the roots opposite to those of Q_∞; hence $M_\infty^{-1} M_\infty \cap Q_\infty = M_\infty \cap Q_\infty$ is contained in

$$Q_\infty \cap Q'_\infty = H_\infty$$

which finishes the proof.

11 Rational points in closed orbits

We will show how Theorem 7 allows to find the following result:

THEOREM 10 (BOREL-HARISH-CHANDRA). *Let V be a vector space defined over k, G a reductive algebraic subgroup of $GL(V)$ defined over k, and ξ a point of V_k whose orbit $G(\xi)$ is closed (as an algebraic variety in V). Then $G_A(\xi) \cap V_k$ has finitely many points modulo G_k.*

(The proof which follows is directly inspired by [2], p. 504–505; we neglect, in this proof, the difficulties due to algebraic geometry, in order to retain only the arithmetic aspect of the problem.)

We will first reduce to the case where G is connected, and we thus choose T, P, M and σ as in the beginning of §10. Denoting by H the stabilizer of ξ in G, the orbit $G(\xi)$ is isomorphic to G/H as algebraic variety over k, thus G/H is affine, and one deduces that H is reductive ([2], p. 499); therefore H_∞ is stable under the involution of G_∞ with respect to a suitable conjugate of M_∞, in other words there is $x \in G_\infty$ such that one has

$$\theta(xH_\infty x^{-1}) = xH_\infty x^{-1} \tag{7}$$

where θ denotes the involution of G_∞ with respect to M_∞. As M_∞ is adapted to T one also has

$$\theta(t) = t^{-1} \tag{8}$$

for $t \in T_\infty^+$.

This said, and observing that, if σ is well chosen, one has

$$G_A = \sigma G_k = G_k\sigma^{-1} = G_k\sigma^{-1}x,$$

it suffices, to establish the theorem, to show that the set

$$\sigma^{-1}x(\xi) \cap V_k$$

is finite, which we will do.

Let thus be $g = mtp \in \sigma$ such that $g^{-1}x(\xi) \in V_k$; one has

$$g^{-1}x(\xi) = \sum_\lambda \xi_\lambda(g)$$

where $\xi_\lambda(g)$ is a rational vector belonging to the weight λ of G with respect to T, thus such that

$$tg^{-1}x(\xi) = \sum \lambda(t)\xi_\lambda(g).$$

But $tg^{-1} = tp^{-1}t^{-1}m^{-1}$ stays in a fixed compact set by Lemma 1 in §10; this also holds for $\sum \lambda(t)\xi_\lambda(g)$ and as $\xi_\lambda(g)$, being rational, is zero or of height larger than a fixed $c > 0$, one deduces from this that, for $\xi_\lambda(g) \neq 0$, one has simultaneously

$$\|\lambda(t)\xi_\lambda(g)\| < 1 \quad \text{and} \quad |\lambda(t)| < 1\,;$$

as $t^2g^{-1}x(\xi) = \sum \lambda(t)^2\xi_\lambda(g)$ this implies that

$$\|t^2g^{-1}x(\xi)\| < 1.$$

But as $t \in T_\infty^+$ the finite part of t^2g^{-1} stays in a fixed compact set and one sees that the infinity component of $t^2g^{-1}x(\xi)$ stays in the intersection of a fixed compact subset of V_∞ with $G_\infty(\xi)$. But as $G(\xi)$ is closed in V from the point of view of algebraic geometry it is clear that $G(\xi)_\infty$ is a closed subvariety of V_∞ from the point of view of real algebraic geometry; moreover, for every $x \in G(\xi)_\infty$, the map $g \to g(x)$ which is a submersion $G \to G(\xi)$ from the point of view of algebraic geometry is a submersion $G_\infty \to G(\xi)_\infty$ from the point of view of real algebraic geometry, so that every orbit

$G_\infty(x)$ is open and thus closed in $G(\xi)_\infty$; hence $G_\infty(\xi)$ is a *closed* subvariety of V_∞ obviously isomorphic to G_∞/H_∞, and as the infinity component of $t^2g^{-1}x(\xi)$ stays in a compact part of V_∞, thus of $G_\infty(\xi) \cong G_\infty/H_\infty$, one sees that the component at infinity of $t^2g^{-1}x$ stays in the product of H_∞ with a fixed compact subset of G_∞, and finally that $t^2g^{-1}x \in C \cdot H_A$ where C is a fixed compact subset of G_A. But one has $t^2g^{-1} = t^2p^{-1}t^{-2}tm^{-1}$ and by using again Lemma 1 of §10 one sees that $tm^{-1}x \in C'H_A$ where C' is compact, thus $tm^{-1} \in C''xH_Ax^{-1}$; using (7) and (8) and applying θ one sees that $t^{-1}m^{-1} \in C''' \cdot xH_Ax^{-1}$, and as

$$g^{-1}x(\xi) = p^{-1}t^{-1}m^{-1}x(\xi) \in C''''H_A(\xi) = C''''(\xi)$$

one sees that $g^{-1}x(\xi)$ stays in the intersection of V_k with a fixed compact subset of V_A, i.e. in a fixed finite set, which finishes the proof.

(In order to supplement Theorem 10 and turning it into a statement of type Hasse-Minkowski, we still have to show — which is easy but not evident — that the elements of $G_A(\xi) \cap V_k$ are the $\eta \in G(\xi)_k$ which are "locally equivalent" to ξ, i.e. which verify $\eta \in G_p(\xi)$ for p finite or not.)

12 Translation to infinity

Theorems 7, 9 and 10 can easily be replaced by statements which include only the situation at infinity, and there are by the way such statements which one finds in [1] and [2].

Suppose that $G \subset GL(V)$ where V is a vector space defined over k, and choose a maximal compact subgroup M_∞ of G_∞ adapted to the torus T, so that one has, after Mostow-Iwasawa, the relation $G_\infty = M_\infty P_\infty^+$ (where P_∞^+ is the connected component of P_∞). It is on the other hand clear that $P_\infty^+ = T_\infty^+ P_\infty^\circ$, where we denote by P_∞° the subgroup of those $p \in P_\infty^+$ for which holds $\chi(p) = 1$ for every rational character χ of P defined over k (denoting by $Z'(T)$ the derived, semi-simple group of $Z(T)$, it is clear that $P_\infty^\circ = Z'(T)_\infty U_\infty$). We call *open Siegel set* in G_∞ every set of the form

$$\sigma = M_\infty T_\infty^+(c)D$$

where D is an open relatively compact subset of P_∞°.

Choose on the other hand for every finite p an open compact subgroup M_p of G_p such that of course, for almost every p, M_p is the subgroup of elements of G_p which preserve a given rational lattice, and let Γ be the subgroup of elements $\gamma \in G_k$ such that one has $\gamma_p \in M_p$ for every finite p (so that Γ is a discrete "arithmetically defined" subgroup of G_∞); it is clear (Theorem 6) that $P_k \backslash G_k / \Gamma$ is finite. This said:

THEOREM 11 (BOREL). *Let (ξ_v) be a finite family of elements of G_k. The following properties are equivalent:*

(a) *there exists an open Siegel set σ in G_∞ such that $G_\infty = \bigcup \sigma\xi_v\Gamma$,*

(b) *one has $G_k = \bigcup P_k\xi_v\Gamma$.*

(One should of course add, to be complete, that if (a) holds, then $\Omega = \bigcup \sigma\xi_v$ is an open fundamental set for Γ in G_∞, i.e. Ω meets only a finite number of $\Omega\gamma$ — but this kind of statement immediately follows from Theorem 9 established above.)

Let us first suppose that $G_k = \bigcup P_k \xi_v \Gamma$ and choose in G_A an open Siegel set σ_A such that $G_A = \sigma_A G_k$; the image σ of σ_A under the projection $G_A \to G_\infty$ is obviously an open Siegel set in G_∞. One has

$$G_A = \bigcup \sigma_A P_k \xi_v \Gamma$$

and consequently every $g \in G_\infty$ can be written as $g = s\pi\xi_v\gamma$ with an $s \in \sigma_A$, a $\pi \in P_k$, a $\gamma \in \Gamma$ and a v suitably chosen; if one denotes generally, for every $x \in G_A$, by x_f the element of G_A deduced from x by replacing the component x_∞ of x by the identity, the relation $g = s\pi\xi_v\gamma$ splits in

$$g = s_\infty(\pi\xi_v\gamma)_\infty, \quad 1 = s_f(\pi\xi_v\gamma)_f; \tag{9}$$

as the "finite" part of an open Siegel set in G_A is compact, the second relation in (9) shows that $(\pi\xi_v\gamma)_f$ stays in fixed compact subset of G_A, thus also $(\pi\xi_v)_f$ since γ_f stays in the product of the M_p, p finite; one deduces from this that, for every v, the $\pi \in P_k$ which appear here stay in a finite number of classes modulo $P_k \cap \Gamma$, or, what amounts to the same, modulo $P_k \cap \xi_v\Gamma\xi_v^{-1}$; thus there are a finite number of elements $\pi_{v\rho} \in P_k$ such that (9) implies $\pi\xi_v \in \pi_{v\rho}\xi_v\Gamma$ for at least one ρ, and as $s_\infty \in \sigma$ the first relation in (9) shows that

$$G_\infty = \bigcup \sigma\pi_{v\rho}\xi_v\Gamma\,;$$

since the union of the $\sigma\pi_{v\rho}$ obviously is an open Siegel set in G_∞ we have proved that (b) implies (a).

Suppose conversely that $G_\infty = \bigcup \sigma\xi_v\Gamma$, and let us show that $G_k = \bigcup P_k\xi_v\Gamma$. Let $\xi \in G_k$; we have

$$\sigma\xi \subset \bigcup \sigma\xi_v\gamma$$

and after Theorem 9 one can find a *finite number* of elements $\gamma_\lambda \in \Gamma$ such that one has

$$\sigma\xi \subset \bigcup \sigma\xi_v\gamma_\lambda.$$

This immediately implies, for a suitable choice of v and λ, that there are elements t_n in $T_\infty^+ \cap \sigma$ which satisfy $t_n\xi \in \sigma\xi_v\gamma_\lambda$ for every n, and $\lim \alpha_i(t_n) = 0$ for every i; as the first relation can also be written as $t_n\gamma \in \sigma$ where

$$\gamma = \xi\gamma_\lambda^{-1}\xi_v^{-1}$$

Lemma 3 of §10 (which applies here since σ is the intersection of G_∞ with an open Siegel set of G_A which, as one can suppose, contains the finite component of γ) shows that for every index j one has either $\alpha_j(t_n) > 1$, which just is not the case, or $\gamma \in P(j)$. In this way one has, for a suitable choice of v and λ, the relation

$$\xi\gamma_\lambda^{-1}\xi_v^{-1} \in \bigcap_{1\le j\le r} P(j)_k = P_k$$

and this shows that $\xi \in P_k\xi_v\Gamma$ which finishes the proof.

Bibliography

[1] BOREL (Armand). Ensembles fondamentaux pour les groupes arithmétiques, Colloque sur la théorie des groupes algébriques [1962 Bruxelles], p. 23–40, Gauthier-Villars, 1962 (Centre belge de Recherches mathématiques).

[2] BOREL (A.) and HARISH-CHANDRA. Arithmetic subgroups of algebraic groups, Annals of Math., Series 2, t. 75, 1962, p. 485–535.

[3] GODEMENT (Roger). Groupes linéaires algébriques sur un corps parfait, Séminaire Bourbaki, t. 13, 1960/61, *n*° 206, 22 p.

[4] MOSTOW (G. D.) and TAMAGAWA (T.). On the compactness of arithmetically defined homogeneous spaces, Annals of Math., Series 2, t. 76, 1962, p. 446–463.

[5] ONO (T.). Sur une propriété arithmétique des groupes commutatifs, Bull. Soc. math., France, t. 85, 1957, p. 307–323.

[6] WEIL (André). Adeles and algebraic groups, Princeton, Institute for advanced Study, 1961 (mimeographed).